DESALINATION TECHNOLOGY

HEALTH AND ENVIRONMENTAL IMPACTS

DESALINATION TECHNOLOGY

HEALTH AND ENVIRONMENTAL IMPACTS

EDITED BY
JOSEPH COTRUVO • NIKOLAY VOUTCHKOV
JOHN FAWELL • PIERRE PAYMENT
DAVID CUNLIFFE • SABINE LATTEMANN

CRC Press is an imprint of the
Taylor & Francis Group, an **informa** business

Co-published by IWA Publishing, Alliance House, 12 Caxton Street, London SW1H 0QS, UK
Tel. +44 (0) 20 7654 5500, Fax +44 (0) 20 7654 5555
publications@iwap.co.uk
www.iwapublishing.com
ISBN 1843393476
ISBN13 9781843393474

CRC Press
Taylor & Francis Group
6000 Broken Sound Parkway NW, Suite 300
Boca Raton, FL 33487-2742

Printed in the United States of America on acid-free paper
10 9 8 7 6 5 4 3 2 1

International Standard Book Number: 978-1-4398-2890-8 (Hardback)

Library of Congress Cataloging-in-Publication Data

Desalination technology : health and environmental impacts / editor, Joseph Cotruvo.
p. cm.
Includes bibliographical references and index.
ISBN 978-1-4398-2890-8 (hardcover : alk. paper)
1. Saline water conversion. I. Cotruvo, Joseph A. II. Title.

TD479.D473 2010
628.1'67--dc22 2010018117

Visit the Taylor & Francis Web site at
http://www.taylorandfrancis.com

and the CRC Press Web site at
http://www.crcpress.com

Contents

List of Figures

List of Tables

Preface

Water is essential to life, and access to sufficient quantities of safe water for drinking and domestic uses and also for commercial and industrial applications is essential for healthful living, enhanced quality of life and well-being, and the opportunity to achieve human and economic development. Many world regions are grossly deficient in the availability of water of sufficient quantity as well as quality. People in many areas of the world have historically suffered from inadequate access to safe water. Some must walk long distances just to obtain sufficient water to sustain life. As a result, they have had to endure health consequences and have not had the opportunity to develop their resources and capabilities to achieve major improvements in their well-being. With growth of the world population, the availability of the limited quantities of fresh water continually decreases.

Most of the world's water is seawater or brackish water, and groundwater that is high in total dissolved solids and either undesirable or unavailable for beneficial uses without the application of technologies capable of removing large portions of the salinity and dissolved solids. Commercial desalination technologies were introduced about 50 years ago and were able to expand access to water, but at high cost. Developments of significant new and improved technologies have now significantly broadened the opportunities to access large quantities of safe water in many parts of the world. Costs are still significant compared with those associated with freshwater sources, but there has been a major cost reduction trend. The desalination option is now much more widely available and probably the principal source of "new" water in the world. Even so, when the alternative is no water or inadequate water quantity for needs and significant harm to health and welfare, greater cost is endurable in many circumstances.

Almost 14,000 desalination plants are in operation throughout the world, producing about 53 million cubic meters of water per day. Facilities for an additional 10.6 million cubic meters of water per day have been contracted. The number is growing rapidly as the need for freshwater supplies grows more acute, technologies improve, and unit costs are reduced.

Desalination plants use waters that are impaired with salts or other contaminants as their sources. It appears that performance, operating, and product quality specifications have evolved virtually on a site-by-site basis relative to source and the specific end-product water use. Most worldwide desalination applications use the World Health Organization's Guidelines for Drinking-Water Quality (GDWQ) in some way as finished water quality specifications. National or state requirements exist in some areas. The GDWQ cover a broad spectrum of contaminants, including inorganic and synthetic organic chemicals, disinfection by-products, microbial indicators, and radionuclides, and are aimed at typical fresh drinking water sources and technologies. Because desalination is applied to nontypical source waters and often uses nontypical technologies, existing national standards and guidelines may not fully cover the unique factors that can be encountered during intake, production, and distribution of desalinated water.

This book addresses drinking water quality, technology, and environmental protection issues in order to assist both proposed and existing desalination facilities to be optimized to ensure that nations and consumers will be able to enjoy the benefits of the expanded access to desalinated water with the assurance of quality, safety, and environmental protection. Apart from the quality and safety of the finished drinking water, numerous other health and environmental protection issues are also evident when considering the impacts of desalination processes, including energy conservation and sustainability. Many of them are not unique to desalination. They may also relate to any large construction project sited in a coastal or other environmentally sensitive area. Protection of the coastal ecosystem and protection of groundwater from contamination by surface disposal of concentrates are examples of issues that must be addressed during the design, construction, and operation of a desalination facility.

Acknowledgments

The support of Dr. Hussein A. Gezairy, World Health Organization (WHO) Regional Director for the Eastern Mediterranean, was determinant for the initiation and development of this critical project. We wish to express special appreciation to Dr. Houssain Abouzaid, Coordinator of the Regional Office for the Eastern Mediterranean Healthy Environments Programme, for initiating the project, to the chairs and members of the several technical committees, and to Dr. Joseph Cotruvo, Technical Advisor, United States, who managed the desalination guidance development process.

The Oversight Committee chaired by Dr. Houssain Abouzaid included Dr. Jamie Bartram, WHO; Dr. Habib El Habr, United Nations Environment Programme/ Regional Office for West Asia; Dr. Abdul Rahman Al Awadi, Regional Organization for the Protection of the Marine Environment; and Dr. Joseph Cotruvo, Technical Advisor.

The Steering Committee chaired by Dr. Houssain Abouzaid consisted of Amer Al-Rabeh, Saudi Arabia; Dr. Anthony Fane, Australia; Dr. Gelia Frederick-van Genderen, Cayman Islands; Dr. Totaro Goto, Japan; Dr. Jose Medina San Juan, Spain; and Kevin Price, United States.

The Technical Working Groups consisted of a balanced group of international expert scientists and engineers with particular expertise in the specialty technical areas. These working groups and their members conducted the scientific analyses and generated the respective chapters that provided the technical content and recommendations.

TECHNOLOGY ENGINEERING AND CHEMISTRY: LARGE AND SMALL FACILITIES

Chair: Nikolay Voutchkov, Water Globe Consulting, LLC, Stamford, Connecticut, USA

Chair: Dr. Corrado Sommariva, ILF Consulting Engineers, Abu Dhabi, UAE

Chair: Mr. Tom Pankratz, Water Desalination Report, Global Water Intelligence, Houston, Texas, USA

Mr. Leon Awerbuch, Leading Edge Technologies, Winchester, Massachusetts, USA

Mr. Nick Carter, Abu Dhabi Regulation and Supervision Bureau, Abu Dhabi, UAE

Dr. Vince Ciccone, Romem Aqua Systems Co. (RASCO) Inc., Woodbridge, Virginia, USA

Mr. David Furukawa, Separation Consultants, Inc., Poway, California, USA

Dr. James Goodrich, National Risk Management Research Laboratory, Environmental Protection Agency, Cincinnati, Ohio, USA
Ms. Lisa Henthorne, Water Standard, Tampa, Florida, USA
Dr. Tom Jennings, Bureau of Reclamation, Washington, DC, USA
Mr. Frank Leitz, Bureau of Reclamation, Denver, Colorado, USA
Mr. John Tonner, Water Consultants International, Mequon, Wisconsin, USA

CHEMICAL ASPECTS OF DESALINATED WATER

Chair: Mr. John Fawell, Consultant, Flackwell Heath, High Wycombe, UK
Chair: Dr. Mahmood Yousif Abdulraheem, Kuwait Foundation for the Advancement of Sciences (KFAS), Kuwait; Advisor, Abu Dhabi Environment Agency, Abu Dhabi, UAE
Dr. Fatimah Al-Awadhi, Kuwait Foundation for the Advancement of Sciences, Kuwait
Dr. Yasumoto Magara, Hokkaido University, Sapporo, Japan
Dr. Choon Nam Ong, National University of Singapore, Singapore

SANITARY MICROBIOLOGY OF PRODUCTION AND DISTRIBUTION OF DESALINATED WATER

Chair: Dr. Pierre Payment, INRS-Institut Armand-Frappier, Laval, Quebec, Canada
Chair: Dr. Michèle Prévost, Ecole Polytechnique de Montréal, Montreal, Quebec, Canada
Chair: Dr. Jean-Claude Block, Laboratory of Physical Chemistry and Microbiology for the Environment, Centre National de la Recherche Scientifique, University Henri Poincaré, Nancy, France
Dr. Henryk Enevoldsen, Intergovernmental Oceanographic Commission Centre on Harmful Algae, University of Copenhagen, Copenhagen, Denmark
Dr. Sunny Jiang, University of California at Irvine, Irvine, California
Dr. Harvey Winters, Fairleigh Dickenson University, Teaneck, New Jersey

MONITORING, SURVEILLANCE, AND REGULATION

Chair: Dr. David Cunliffe, Environmental Health Service, Department of Health, Adelaide, South Australia, Australia
Dr. Marie-Marguerite Bourbigot, Veolia Environment, Paris
Dr. Shoichi Kunikane, Department of Water Supply Engineering, National Institute of Public Health, Wako, Japan
Dr. Richard Sakaji, East Bay Municipal Utility District, Oakland, California

ENVIRONMENTAL EFFECTS AND IMPACT ASSESSMENTS

Chair: Dr. Sabine Lattemann, Federal Environment Agency (UBA), Berlin
Mr. Bradley Damitz, Monterey, California

Dr. Klaus Genthner, Bremen, Germany
Dr. Hosny Khordagui, United Nations Economic and Social Commission for Western Asia, Beirut
Dr. Greg Leslie, University of New South Wales, Kennington, Australia
Dr. Khalil H. Mancy, Emeritus, University of Michigan, Ann Arbor, Michigan, USA
Mr. John Ruettan, Resource Trends, Escondido, California
Dr. Samia Galal Saad, High Institute of Public Health, Alexandria, Egypt

This book was originally commissioned by the WHO Regional Office for the Eastern Mediterranean. It presents a comprehensive overview of the subject, including technologies, health and environmental impact assessment, and it will provide a basis for a desalination and health publication that WHO will produce in the coming year. We especially wish to acknowledge the organizations that generously sponsored the development process. These included the AGFUND, the Kuwait Foundation for the Advancement of Sciences (KFAS), the U.S. Environmental Protection Agency's National Risk Management Research Laboratory (Cincinnati, Ohio), the American Water Works Association Research Foundation (Denver, Colorado)/ Water Research Foundation, the Water Authority of the Cayman Islands, the U.S. Bureau of Reclamation (Denver, Colorado), and the National Water Research Institute (Fountain Valley, California). We gratefully acknowledge the expertise and in-kind services provided by all of the expert participants without which this book would not have been possible.

This final report was developed partly under Cooperative Agreement No. CR831028 awarded by the U.S. Environmental Protection Agency. EPA made comments and suggestions on the document intended to improve the scientific analysis and technical accuracy of the document. These comments are included in the report. However, the views expressed in this document are those of the authors and EPA does not endorse any products or commercial services mentioned in this publication.

Abbreviations and Acronyms

ADI: Acceptable daily intake
AI: Aggressiveness index
AM: Anion transfer membrane
AOC: Assimilable organic carbon
BOD: Biochemical oxygen demand
BOM: Biodegradable organic matter
BTEX: Benzene, toluene, ethylbenzene, xylenes
BWRO: Brackish water reverse osmosis
CCPP: Calcium carbonate precipitation potential
CEB: Chemically enhanced backwash
CM: Cation transfer membrane
CT: Concentration × time
CVD: Cardiovascular disease
DBP: Disinfection by-product
ED: Electrodialysis
EDR: Electrodialysis reversal
EDTA: Ethylenediaminetetraacetic acid
EIA: Environmental impact assessment
GDWQ: WHO Guidelines for Drinking-Water Quality
HDD: Horizontal directionally drilled
HDPE: High-density polyethylene
HPC: Heterotrophic plate count
LR: Larson ratio
LSI: Langelier saturation index
MED: Multiple effect distillation
MED-TC: Multiple effect distillation system with a thermocompressor device
MF: Microfiltration
MS2: Male-specific bacteriophage (F-RNA)
MSF: Multistage flash distillation
MVC: Mechanical vapor compression
NCG: Noncondensable gas
NDMA: Nitrosodimethyl amine; dimethylnitrosamine
NF: Nanofiltration
NGO: Nongovernmental organization
NOM: Natural organic matter
ORP: Oxidation–reduction potential
PR: Performance ratio
PVC: Polyvinyl chloride

RO: Reverse osmosis
SDI: Silt density index
SWRO: Seawater reverse osmosis
TBT: Top brine temperature
TDS: Total dissolved solids
THM: Trihalomethane
TOC: Total organic carbon
ToR: Terms of reference
TSS: Total suspended solids
TTHM: Total trihalomethanes
UF: Ultrafiltration
UN: United Nations
USA: United States of America
UV: Ultraviolet
VCD: Vapor compression distillation
WHO: World Health Organization
WSP: Water safety plan
ZID: Zone of initial dilution

The Editors

Joseph A. Cotruvo, Ph.D., in his career has run the gamut from basic water chemistry and technology, federal and international regulatory development, health guidance, research into water science, technology and toxicology, and water and wastewater supply oversight. He is president of Joseph Cotruvo & Associates LLC. His doctorate is in physical organic chemistry from Ohio State University. He was the first director of the U.S. Environmental Protection Agency's (USEPA) Drinking Water Criteria and Standards Division, and also director of USEPA's Risk Assessment Division in the Office of Pollution Prevention and Toxics. He developed much of the methodology for U.S. drinking water standards, and applied it initially to the THM regulations. He has been a core member of the World Health Organization's Steering Committee developing the Guidelines for Drinking Water Quality and has engaged in numerous projects producing major water and health policy products for WHO including zoonoses, heterotrophic plate counts, and mineral nutrients in drinking water, aircraft water sanitation, and food process disinfection. He was overall editor and contributor to the WHO/WPC monograph *Health Aspects of Plumbing*, and he prepared WHO's Calcium and Magnesium in Drinking Water: Public Health Significance. He has worked on desalination-related issues in Kuwait and Abu Dhabi, and he is a member of the Technical Committee on National Drinking Water Quality Standards in Singapore.

Dr. Cotruvo has been principal investigator (PI) and co-principal investigator on numerous projects investigating drinking water safety, technology, security, decontamination, nontraditional techniques for water provision, and disinfection by-products formation and health issues. Currently he is PI on a toxicological study to elucidate the bromate risk, if any, at low drinking water doses. He is a member of the Research Advisory Boards of the WateReuse Foundation and the National Water Research Institute and serves on the Health Advisory Board for the Orange County (California) Groundwater Replenishment System, and the San Diego Water Reuse Science Advisory Panel. He is an honorary life member of AWWA, and also a member of the Retail Services (Water Quality) Committee of the board of directors of the District of Columbia Water and Sewer Authority.

Pierre Payment, Ph.D., obtained his M.Sc. (microbiology and immunology) in 1971 and his Ph.D. (microbiology and immunology) in 1974 from the University of Montreal. His postdoctoral studies were done at the Baylor College of Medicine in Houston (Texas). It is during that period that he became interested in the problems associated with the presence of microorganisms in water. Returning to Montreal in 1975, he became professor at Institut Armand-Frappier. He is now a full professor at this institution and has been very active both in clinical microbiology and public health. He has been head of the Electron Microscopy Laboratory and head of the Veterinary Virology Diagnostic Services. Apart from his activities as a researcher, he was director of the Technology Transfer Unit at INRS-Institut Armand-Frappier and director of the Intellectual Property Management Network of the University of Quẻbec. He was also president of the Canadian Association of Clinical Microbiology and Infectious Diseases (CACMID) and the organizer of their annual meeting. He is knowledgable on many aspects of water treatment and microbiology and his current research activities are centered on the health effects of drinking water. As an expert, he has participated in activities of the USEPA, WHO, Health Canada, OECD, the Walkerton Inquiry and was a member of the Advisory Scientific Committee of the Joint International Commission (Great Lakes).

John Fawell comes from a toxicology background but has worked on drinking water quality issues for nearly 30 years. He has the following degrees from Bath University (UK): B.Sc. (Hons), CBiol, MIBiol, MCIWEM, Dip RCPATH. He has carried out research on a wide range of water contaminants and he was chief scientist at the U.K. Water Research Centre's National Centre for Environmental Toxicology and is now an independent consultant. He works on drinking water standards and guidelines among a range of other topics and has been one of the coordinators of the WHO guidelines for drinking water quality since 1988. He is closely involved in the implementation of water safety plans and has also worked on the quality issues associated with new water sources including desalination.

Sabine Lattemann, Ph.D., is a marine scientist specialized in environmental impact assessment (EIA) studies, particularly of seawater desalination, offshore wind energy development projects, and maritime shipping impacts. She is a scientific officer at the German Federal Environmental Protection Agency (Umweltbundesamt), and chaired the environmental working group of the World Health Organization Project "Desalination for safe water supply" from 2004 to 2007. From 2007 to 2009, she carried out her Ph.D. within the EU-funded research project "Membrane-based desalination, an integrated approach" at the University of Oldenburg, Germany, and the UNESCO-IHE Institute for Water Education in Delft, The Netherlands. Sabine is the main author of the book *Seawater Desalination: Impacts of Brine and Chemical Discharges on the Marine Environment* (Balaban Desalination Publications, 2003) and of *Resources and Guidance Manual for Environmental Impact Assessment of Desalination Projects* published by the United Nations Environment Programme in 2008.

David Cunliffe, Ph.D., received a B.Sc. with honours in microbiology from the University of Adelaide and a Doctor of Philosphy from Flinders University. Dr. Cunliffe is the Principal Water Quality Adviser with the South Australian Department of Health. He is a regulator with over 25 years experience in dealing with public health aspects of drinking water and recycled water.

Dr. Cunliffe has contributed to a range of national and international guidelines including World Health Organization (WHO) and Australian guidelines for drinking water quality and use of recycled water. He has published on a wide range of issues relating to water quality, rainwater, and the management of drinking water and recycled water supplies. He is a member of the WHO Drinking Water Quality Committee and the Australian Water Quality Advisory Committee. Dr. Cunliffe has received the Public Service Medal and the Premier's Water Medal for services to water quality in Australia.

Nikolay Voutchkov has over 25 years experience in the field of seawater desalination and water treatment. Currently, he is an independent expert providing desalination project advisory services through his company, Water Globe Consulting, LLC. He has PE and BCEE degrees from the Technical University of Civil Engineering and Architecture, Sofia, Bulgaria. Mr. Voutchkov has published over 30 technical articles and co-authored several books on desalination, including technology and design guidelines for the American Water Works Association and the Australian Water Association. He is a registered professional engineer and a Board Certified Environmental Engineer by the American Academy of Environmental Engineers. Mr. Voutchkov serves on the Research Advisory Committee of the WateReuse Research Foundation and the Research Project Advisory Committees for the American Water Works Association and the Water Environment Federation. Mr. Voutchkov's work in the field of desalination research and technology has been recognized with numerous awards, including the 2006 Global Grand Prize for Innovation by the International Water Association.

Contributors

Mahmood Yousif Abdulraheem
Kuwait Foundation for the Advancement of Sciences
Kuwait
Abu Dhabi Environment Authority
Abu Dhabi, UAE

Houssain Abouzaid
Regional Office for the Eastern Mediterranean Healthy Environments Programme

Fatimah Al-Awadhi
Kuwait Foundation for the Advancement of Sciences
Kuwait

Jean-Claude Block
Laboratory of Physical Chemistry and Microbiology for the Environment
Centre National de la Recherche Scientifique
University Henri Poincaré
Nancy, France

Nick Carter
Abu Dhabi Regulation and Supervision Bureau
Abu Dhabi, UAE

Bradley Damitz
Monterey, CA

Hosny Khordagui
United Nations Economic and Social Commission for Western Asia
Beirut, Lebanon

Shoichi Kunikane
Department of Water Supply Engineering
National Institute of Public Health
Wako, Japan

Greg Leslie
University of New South Wales
Kennington, Australia

Yasumoto Magara
Hokkaido University
Sapporo, Japan

Khalil H. Mancy
University of Michigan
Ann Arbor, MI

Choon Nam Ong
National University of Singapore
Singapore

Tom Pankratz
Water Desalination Report
Global Water Intelligence
Houston, TX

Michèle Prévost
Ecole Polytechnique de Montréal
Montreal, Quebec, Canada

Richard Sakaji
East Bay Municipal Utility District
Oakland, CA

Corrado Sommariva
Texas ILF Consulting Engineers
Abu Dhabi, UAE

John Tonner
Water Consultants International
Mequon, WI

1 Overview of the Health and Environmental Impacts of Desalination Technology*

Joseph Cotruvo and Houssain Abouzaid

CONTENTS

Water is essential to life, and access to sufficient quantities of safe water for drinking and domestic uses and for commercial and industrial applications is essential for healthful living, enhanced quality of life and well-being, and the opportunity to achieve human and economic development. Many world regions are grossly deficient

* This chapter was derived in part from an assessment prepared by J.A. Cotruvo for the World Health Organization's Regional Office for the Eastern Mediterranean. The assessment was also the basis for a paper published in the Water Conditioning & Purification Magazine (Cotruvo 2005) and a paper presented at the Environment 2007 International Conference on Integrated Sustainable Energy Resources in Arid Regions in Abu Dhabi (Cotruvo and Abouzaid 2007).

in availability of water of sufficient quantity as well as quality. People in many areas of the world have historically suffered from inadequate access to safe water. Some must walk long distances just to obtain sufficient water for basic needs. As a result, they have had to endure adverse health consequences and have not had the opportunity to develop their resources and capabilities to improve their lot. With current world population growth, the availability of limited quantities of fresh water is continually decreasing.

One of the World Health Organization's (WHO's) principal missions is to improve access to sufficient quantities of safe water throughout the world. The United Nations has declared 2005–2015 the International Decade for Action "Water for Life," setting a world agenda that focuses increased attention on water-related issues. This initiative is of extraordinary importance in a world where preventable diseases related to water and sanitation claim the lives of about 3.1 million people a year, most of them children less than 5 years old. Of these, about 1.6 million people die from diarrheal diseases from lack of safe drinking water and adequate sanitation.

By including drinking water supply, sanitation, and hygiene in the Millennium Development Goals, the world community has acknowledged the importance of their promotion as development and health interventions and has set a series of goals and targets accordingly. Target 7c of Goal 7 (Ensure environmental sustainability) requests the world to "halve by 2015 the proportion of people without sustainable access to safe drinking-water and basic sanitation." The task is huge: in 2002, 1.1 billion people (two-thirds of them in Asia and 42% of the population in sub-Saharan Africa) lacked access to improved water sources. At least 2.6 billion people lack access to improved sanitation; over half of them live in China and India. Only 31% of rural inhabitants in developing countries have access to improved sanitation, compared with 73% of urban dwellers. Achieving the Millennium Development Goal drinking water and sanitation target requires that 97 million additional people gain access to drinking water services and 138 million additional people gain access to sanitation annually up to 2015.

1.1 DESALINATION BACKGROUND

Most of the world's water is seawater, brackish water, and groundwater that is high in total dissolved solids (TDS). These waters are either undesirable or unavailable for use without the application of technologies capable of removing large portions of the salinity and dissolved solids. Desalination technologies were introduced, especially in the West Asia Gulf region, about 50 years ago, and they were able to expand access to water, but at relatively high costs. Development of significant new and improved technologies has now greatly broadened the opportunities to access large quantities of safe water in many parts of the world. Costs are still significant compared with those associated with freshwater sources, but there has been a major cost reduction trend, and the desalination option is now much more widely available and, along with water reuse, the principal source of "new" water in the world. Higher energy costs have stopped the overall price reductions, although the efficiency of the technology continues to improve. Even so, when the alternative is no water or inadequate water quantity for needs and when significant harm to health and welfare can occur, greater cost is tolerable in many circumstances.

Desalination of seawater and brackish water along with planned wastewater reuse for indirect potable and nonpotable applications (e.g., irrigation) are the world's sources of "new" water, and they have been growing rapidly worldwide in recent years. This is because the need to produce more water and efficiently use water to satisfy the needs of growing and more demanding populations has become acute. These technologies are more complex than the more traditional technologies usually applied to relatively good-quality fresh waters. As such, the cost of production is greater than that from freshwater sources, but the technologies are being applied in areas where the need is also greater. Desalination and water reuse share some of the same technologies, so the science and technology of both processes have developed somewhat in tandem. Indeed, it is appropriate to include reuse with desalinated water to improve the efficiency of the total. This book focuses upon desalination and examines the major technologies and the additional health and environmental considerations involved relative to those associated with water production from more traditional sources.

As of 2007, almost 14,000 desalination plants were in operation throughout the world, producing about 53 million cubic meters of water per day, and an additional 10.6 million cubic meters of capacity are contracted. Capacity grew by 24.5% in 2007 alone (WDR 2008). About 50% of the capacity exists in the West Asia Gulf region. North America has about 17%, Asia (apart from the Gulf) about 10%, North Africa about 8%, Europe about 7%, and Australia a bit over 1% (GWI and IDA 2006), but this distribution is changing as more demand develops in Australia, Europe, and North America. Capacity is expected to reach at least 94 million cubic meters per day by 2015 (WDR 2006). Desalination plant capacities range from more than 1 million cubic meters per day down to 20–100 m^3/day. Home-sized reverse osmosis units may produce only a few liters per day, but these are usually applied to fresh waters with higher than usual TDS content. Over the next 10 years, at least $100 billion for desalination is projected to be needed in the Arabian Gulf States alone just to keep up with economic growth and water demand, according to a 2006 report (EMS 2006).

Any large-scale facility brings with it the potential for unintended collateral health and environmental upsets unless adequate consideration is given to these concerns at the time of planning and before construction and operation. Many of these questions are not unique to desalination, and some of them are also dealt with in the WHO *Guidelines for Drinking-water Quality* (GDWQ) (WHO 2004):

- What are the principal technologies for desalination, and what factors influence their selection and performance?
- How can the aggressiveness of desalinated water and its corrosivity be managed?
- What quality management guidance is available for blending waters that are added postdesalination for adjustment and stabilization?
- What is the appropriate guidance for aesthetic and stability factors (e.g., TDS, pH, taste and odor, turbidity, corrosion indices)?

- Should optimal finished water reflect potentially nutritionally desirable components of reconstituted finished water that are removed during treatment (e.g., calcium, magnesium, fluoride)?
- What are the appropriate quality specifications for the safety of chemicals and materials used in production and in contact with the water (e.g., coagulants, disinfectants, pipes and surfaces in desalination plants, distribution systems)?
- How should monitoring of plant performance and water during distribution be designed (e.g., key chemicals and microbiological parameters and monitoring frequencies)?
- What are appropriate approaches for environmental impact assessments and protection factors relating to siting, marine ecology, groundwater protection, energy production, and air quality?

Finally, the overall management of the operation of any water system, including desalination, is addressed in the use of water safety plans (WSPs). The most effective way to consistently ensure the safety of a drinking water supply is through the use of a comprehensive planning, risk assessment, and risk management approach that encompasses all of the steps in the water supply train from the catchment to the consumer. WHO has developed the systematized WSP approach based upon the understanding of water system function derived from the worldwide history of successful practices for managing drinking water quality (WHO 2005a, 2008). The WSP concept draws upon principles and concepts of sanitary surveys, prevention, multiple barriers, vulnerability assessments and quality management systems such as the hazard assessment and critical control point approach used in the food industry. A WSP has three key components: (1) system assessment, (2) measures to monitor and control identified risks, and (3) management plans describing actions to be taken during normal operations or incident conditions. These are guided by health-based targets (drinking water standards, guidelines, and codes) and overseen through surveillance of every significant aspect of the drinking water system.

This introductory chapter reviews the issues associated with the use of desalinated water. It briefly describes the technologies and management approaches, water quality issues, and health and environmental topics that are addressed in detail in the book:

- *Water quality, technology, and health issues*: Drinking water systems should be designed to produce and deliver to consumers safe drinking water that meets all health and aesthetic quality specifications. Several technologies have applications for producing high-quality water from nonfreshwater sources. The WHO GDWQ provides comprehensive information in this respect, and the guidelines apply equally to conventional and desalinated drinking water. As a result of its nature, origin, and typical locations where practiced, desalinated water involves some additional issues to be considered in respect to both potential chemical and microbial components. This document addresses both types of contaminants, and it additionally reflects some issues that are specific to desalinated drinking water.

- *Aesthetics and water stability*: Although not directly health related, aesthetic factors such as taste, odor, turbidity, and TDS affect palatability and thus consumer acceptance of water and, indirectly, health. Corrosion, hardness, and pH have economic consequences, and they also affect the leaching of metals and other pipe components during distribution. Chemical additives and blending are used to adjust these parameters and the composition and lifespan of the distribution network intimately related to them.
- *Blending waters*: Blending is used to increase the TDS and improve the stability of finished desalinated water. Microbial components of the blending water can also affect the quality and safety of the finished water, because the blended water may not receive any further treatment beyond residual disinfection. Some contaminants may be best controlled by selection or pretreatment of the blending water. Some of the microorganisms in the blending water could be resistant to the residual disinfectant, contribute to biofilms, or be inadequately represented by surrogate microbial quality measurements such as *Escherichia coli* or heterotrophic plate counts (HPCs).
- *Nutritionally desirable components*: Desalinated water is stabilized by adding lime and other chemicals. The diet is the principal source of nutrient minerals that are removed by water treatment, but there is a legitimate question as to the optimal mineral balance of drinking water to ensure quality and health benefits. There is a consensus that dietary calcium and magnesium are important health factors, as well as certain trace metals, and fluoride is also considered to be beneficial for dental and possibly skeletal health by most authorities (WHO 2005b, 2006; Cotruvo 2006). There is a public health policy and practical economic question as to whether and to what extent drinking water should be reconstituted with removed nutritional elements, and this question would be addressed differently in different dietary, political, and social environments. It should be considered as part of the health impact assessment within the environmental impact assessment (EIA).
- *Chemicals and materials used in water production*: Chemicals used in desalination processes are similar to those used in standard water production; however, they may be used in greater amounts and under different conditions. Polymeric membranes as used in desalination processes are not yet as widely used in conventional water treatment, and metals and other components could be subjected to greater than usual thermal and corrosion stresses in desalination compared with conventional water treatment and distribution. The WHO GDWQ address chemicals and materials used in drinking water treatment. The guidelines provide some recommended quality and dose specifications, but they also encourage the application of guidelines for the quality and safety of direct and indirect additives by national and international institutions. This book encourages governments to establish systems for specifying the appropriateness and quality of additives encountered in desalination, or rather, more efficiently and to avoid unnecessary duplication, to adopt existing credible recognized standards and accept product quality certification from recognized organizations for products appropriate for specific desalination conditions.

- *Water quality and distribution system monitoring*: Desalination processes utilize nontraditional water sources and technologies and produce drinking water that is different from that provided from usual sources and by usual processes. Desalination may also be practiced in locations with warm climates and longer distribution networks. Recommendations for monitoring for process surveillance and distributed water quality are provided to assist water suppliers and regulators. Some chemical and microbial monitoring for desalination system management and to ensure safety should be site specific. Operators and authorities should consider identifying a small number of key parameters associated with source and process, included in or in addition to the WHO GDWQ, that would be useful in particular circumstances, as well as articulating basic principles that should be utilized when designing a monitoring scheme.
- *Environmental quality and environmental impact assessments*: As with any major project, large-scale desalination projects can have significant effects on the environment during construction and operation. Procedures and elements of EIA are provided in this guidance to assist project designers and decision makers and regulators to anticipate and address environmental concerns that should be considered when undertaking a project. Among the factors that are addressed are siting considerations, coastal zone/marine protection regarding withdrawal and discharge, air pollution from energy production and consumption, groundwater protection from drying beds, and leachates and sludge disposal. An expanded discussion of EIA issues by the same author beyond what is found in Chapter 6 of this book has been published by UNEP (2008).

Desalination of seawater and brackish waters is a highly developed and integrated set of processes that adds several dimensions of complexity beyond what is typically involved in the production of drinking water from freshwater sources. This chapter provides insights into the concept of drinking water production and treatment and the elements that are managed in that process, as well as integrated health and environmental concerns and management approaches for ensuring the quality and safety of drinking water at the consumer's tap.

1.2 WATER PRODUCTION ISSUES SPECIFIC TO DESALINATION

Drinking water production processes can be divided into three broad categories, each of which will impact the quality of the finished water received by the consumer: (1) source water, (2) treatment technology, and (3) distribution system management. Some of the factors and issues that distinguish desalination from most typical drinking water operations are as follows:

- Source water
  - TDS in the range of about 5,000–45,000 mg/L
  - High levels of particular ions, including sodium, chloride, calcium, magnesium, bromide, iodide, and sulfate
  - Total organic carbon (TOC) type

    - Potential for petroleum contamination
    - Microbial contaminants and other organisms
- Treatment technology
    - Reverse osmosis (RO) membranes
    - Thermal processes
    - Leachates from system components
    - Pretreatment and antifouling additives
    - Disinfection by-products (DBPs)
    - Posttreatment blending with source waters
- Distribution system management
    - Corrosion control additives
    - Corrosion products
    - Bacterial regrowth, including nonpathogenic HPC and pathogens, such as *Legionella*

Related issues that needed to be considered include

- Whether there are risks from consumption of water with low TDS either from general reduced mineralization or reduced dietary consumption of specific minerals or from corrosivity toward components of the plumbing and distribution system
- Environmental impacts of desalination operations and brine disposal
- Whether microorganisms unique to saline waters may not be removed by the desalination process or postblending disinfection
- Monitoring of source water, process performance, finished water, and distributed water to ensure consistent quality at the consumer's tap

1.2.1 Source Water Composition

A typical freshwater source for producing potable water could be a river, lake, impoundment, or shallow or deep groundwater. The water could be virtually pristine, affected by natural contaminants or impacted by agricultural and anthropogenic waste discharges. Even a pristine source may not be wholly desirable, because it could contain excess minerals and suspended particulates that adversely affect the taste and aesthetic quality or safety, as well as natural organic materials (TOC and total organic nitrogen) that could adversely affect the quality of the finished water and place demands on the treatment processes. The range of mineralization of most fresh waters considered to be aesthetically desirable could be from less than 100 mg/L to about 500 mg/L. Microbial contamination can occur in even pristine source water, but especially in surface waters. Many surface waters are significantly impacted by treated or untreated sewage discharges, agricultural or industrial wastes, and surface runoff, so virtually all surface waters require filtration and disinfection prior to becoming acceptable drinking water.

Groundwaters often benefit from natural filtration from passage through the ground and underground storage, but they can also be naturally contaminated (e.g., TDS, arsenic, or excess fluoride). If they are "under the influence of surface water," they

TABLE 1.1
Major Ion Composition of Seawater

	Concentration (mg/L)			
Constituent	**Normal Seawater**	**Eastern Mediterranean**	**Arabian Gulf at Kuwait**	**Red Sea at Jeddah**
Chloride (Cl^-)	18,980	21,200	23,000	22,219
Sodium (Na^+)	10,556	11,800	15,850	14,255
Sulfate (SO_4^{2-})	2,649	2,950	3,200	3,078
Magnesium (Mg^{2+})	1,262	1,403	1,765	742
Calcium (Ca^{2+})	400	423	500	225
Potassium (K^+)	380	463	460	210
Bicarbonate (HCO_3^-)	140	—	142	146
Strontium (Sr^{2+})	13	—	—	—
Bromide (Br^-)	65	155	80	72
Boric acid (H_3BO_3)	26	72	—	—
Fluoride (F^-)	1	—	—	—
Silicate (SiO_3^{2-})	1	—	1.5	—
Iodide (I^-)	<1	2	—	—
Other	1	—	—	—
Total dissolved solids	34,483	38,600	45,000	41,000

Note: — = not reported.
Source: From I.S. Al Mutaz, Water desalination in the Arabian Gulf Region, in *Water management, purification and conservation in arid climates. Vol. 2. Water purification*, ed. M.F.A. Goosen and W.H. Shayya (Lancaster: Technomic Publishing Company, 2000), 245–65.

can also become contaminated by surface discharges of sewage, and agricultural and industrial waste or spills, particularly if the aquifer is shallow and the overlaying soil is porous and does not retard migration of some contaminants or disposal in unlined ponds is practiced. However, many groundwaters are sufficiently protected that they may be consumed without further treatment, or possibly only disinfection.

Seawaters and brackish waters, however, are defined by the extent of the mineralization that they contain. Thus, their composition includes substantial quantities of minerals, which is partly a function of their geographic location. They also contain organic carbon and microbial contaminants, and they can also be impacted by waste discharges. Tables 1.1 and 1.2 provide information on the typical mineral composition of several seawaters. Obviously, special technologies will be required to convert these waters into drinking water that would be safe and desirable to consume.

1.2.2 Treatment Technologies

Treatment of fresh, low-salinity waters centers on particulate removal and microbial inactivation. Thus, filtration and disinfection are the conventional technologies

TABLE 1.2
Major Ion Composition of a Raw Brackish Water

Constituent	Design Value (mg/L)	Design Range (mg/L)
Calcium (Ca^{2+})	258	230–272
Magnesium (Mg^{2+})	90	86–108
Sodium (Na^{+})	739	552–739
Potassium (K^{+})	9	NK
Strontium (Sr^{2+})	3	NK
Iron (Fe^{2+})	<1	0–<1
Manganese (Mn^{2+})	1	0–1
Bicarbonate (HCO_3^-)	385	353–385
Chloride (Cl^-)	870	605–888
Sulfate (SO_4^{2-})	1011	943–1208
Nitrate (NO_3^-)	1	NK
Phosphate (PO_4^{3-})	<1	NK
Silica (SiO_2)	25	NK
Total dissolved solids	3394	2849–3450
pH	8.0	7.8–8.3
Temperature	42°C	36°C–47°C

Note: NK = not known: the sum of these ions is estimated to be between 30 and 40 mg/L.

Source: United States Bureau of Reclamation, Solicitation No. DS-7186 (Denver: U.S. Department of the Interior, 1976), 33.

used, with some exceptions. Coagulation, sedimentation, and rapid sand filtration are commonly used in surface waters, with chlorine or chlorine dioxide and possibly ultraviolet (UV) light for primary disinfection and sometimes chloramines for secondary disinfection. Ozone is used for several purposes. Some reduction of natural organics occurs in the coagulation/filtration process. Microfiltration and ultrafiltration are becoming more widely used, and powdered activated carbon and granular activated carbon are used for taste and odor and sometimes to reduce organics. Softening is sometimes practiced to reduce hardness caused by calcium and magnesium. Targeted technologies are used in some applications, such as for arsenic or nitrate removals.

Home treatment technologies are usually applied as a polish on supplied water; however, there are some technologies that can provide complete treatment and purification. Those technologies should be tested under rigorous conditions and certified by a credible independent organization to meet the claims that they make. They can be at the point of use (end of tap) or point of entry (whole house), which treat all of the water entering the home. The most common systems usually involve ion exchange water softening, or activated carbon for chlorine taste and some organics or iron removal. There are also technologies available for controlling specific classes of contaminants, and some can be used to meet standards and guidelines for various chemicals and microbial contaminants. Disinfection techniques are also available for home or traveler use.

In contrast, desalination processes remove dissolved salts and other materials from seawater and brackish water. Related membrane processes are also used for water softening and wastewater reclamation. The principal desalination technologies in use are distillation and membrane technologies. Desalination technologies are energy intensive, and research is continually evolving approaches that improve efficiency and reduce energy consumption. Cogeneration facilities are now the norm for desalination projects.

1.2.2.1 Distillation/Thermal Technologies

The principal distillation systems include multistage flash distillation (MSF), multi-effect distillation (MED), and vapor compression distillation (VCD). Distillation plants can produce water with TDS in the range 1–50 mg/L. Alkaline cleaners remove organic fouling, and acid cleaners remove scale and salts.

A simplified outline of the process is provided in Figure 1.1. In distillation processes, source water is heated and vaporized; the condensed vapor has very low TDS, whereas concentrated brine is produced as a residual. Inorganic salts and many high-molecular-weight natural organics are nonvolatile and thus easily separated; however, there are circumstances in which volatile petroleum chemicals are present due to spills and other contamination. Even though the vapor pressures of petrochemicals can range from low to very high, some of those with higher molecular weights can also be steam-distilled.

For water, the boiling point (where the vapor pressure of the liquid is the same as the external pressure) is 100°C at 101.3 kPa. As the concentration of solute increases, the boiling point of the solution increases; as the pressure is decreased, the boiling temperature decreases. The amount of energy required to vaporize a liquid is called the heat of vaporization. For water, this amounts to 2256 kJ/kg at 100°C. The same amount of heat must be removed from the vapor to condense it back to liquid at the boiling point. In desalination processes, the heat generated from vapor condensation is transferred to feedwater to cause its vaporization and thus improve the thermal efficiency of the process and reduce fuel consumption and cost.

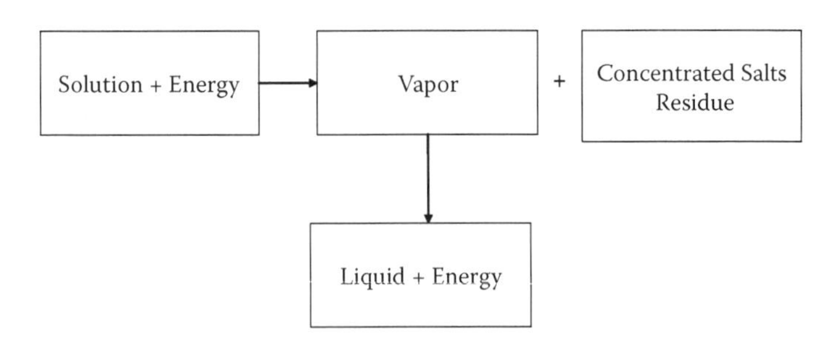

FIGURE 1.1 Distillation process representation.

1.2.2.1.1 Multistage Flash Distillation

MSF plants are major contributors to the world desalting capacity. The principle of MSF is that heated water will boil rapidly (flash) when the pressure of the vapor is rapidly reduced below the vapor pressure of the liquid at that temperature. The vapor that is generated is condensed onto surfaces that are in contact with feedwater, thus heating it prior to its introduction into the flash chamber. This will recover most of the heat of vaporization. Approximately 25%–50% of the flow is recovered as fresh water in multistage plants. Characteristics of MSF plants include high feedwater volume and flow, corrosion and scaling in the plant, and high rates of use of treatment chemicals.

1.2.2.1.2 Multiple Effect Distillation

Configurations of MED plants include vertical or horizontal tubes. Steam is condensed on one side of a tube, with heat transfer causing evaporation of saline water on the other side. Pressure is reduced sequentially in each effect (stage) as the temperature declines, and additional heat is provided in each stage to improve performance.

1.2.2.1.3 Vapor Compression Distillation

VCD systems function by compressing water vapor, causing condensation on a heat transfer surface (tube); this allows the heat of condensation to be transported to brine on the other side of the surface, resulting in vaporization. The compressor is the principal energy requirement. The compressor increases the pressure on the vapor side and lowers the pressure on the feedwater brine side to lower its boiling temperature.

1.2.2.2 Membrane Technologies

Common membranes are polymeric materials such as (originally) cellulose triacetate or more likely polyamides and polysulfones. Membranes are typically layered or thin-film composites. The surface contact layer (rejection layer) is adhered to a porous support, which can be produced from the same material as the surface. Membrane thickness is on the order of 0.05 mm. Selection factors for membranes include pH stability, working life, mechanical strength, pressurization capacity, and selectivity and efficiency for removal of solutes. Membranes are located in a module, and they can be configured as hollow fiber, spiral, plate, and tubular. Each has its own characteristics that affect selection in particular cases. Hollow fiber and spiral configurations generally have more favorable operating characteristics of performance relative to cost, and they are most commonly used. Operating pressures are in the range of 1700–6900 kPa. Membranes used for electrodialysis are 0.13–1.0 mm (typically 0.5 mm) polymeric materials assembled in plate- and frame-type stacks. These membranes operate at feed pressure of 70–700 kPa and are oxidation resistant.

There are numerous compositions of membranes within each category. Table 1.3 provides some generalized performance expectations for four major categories of membrane systems. The larger-pore membrane systems, such as microfiltration (MF) and ultrafiltration (UF), are often used as pretreatments to remove larger particulate contaminants and to reduce the loadings on the more restrictive membranes (e.g., RO) and extend their performance and run times.

TABLE 1.3
Comparison of Membrane Process Performance Characteristics

Membrane Type	Nominal Pore Size (μm) (approximate)	Constituents Removed
Microfiltration	0.1–1	Particulates, bacteria, protozoa
Ultrafiltration	0.001–0.1	Viruses, large and high-molecular-weight organics (e.g., pyrogens)
Nanofiltration	±0.001	Multivalent metal ions, some organics
Reverse osmosis	0.0001–0.001	Seawater and brackish water desalination, salts and organics larger than about 100–300 Da

Source: From http://www.watertreatmentguide.com.

1.2.2.2.1 Reverse Osmosis

RO systems reverse the natural process of solvent transport across a semipermeable membrane from a region of lower solute concentration into one of higher solute concentration to equalize the free energies. In RO, external pressure is applied to the high-solute (concentrated) water to cause solvent (water) to migrate through the membrane, leaving the solute (salts and other nonpermeates) in a more concentrated brine. Some membranes will reject up to 99% of all ionic solids and commonly have molecular weight cutoffs in the range of 100–300 Da for organic chemicals. Increased pressure increases the rate of permeation; however, fouling would also increase. Figure 1.2 illustrates the basic RO process, which includes pretreatment, membrane transport, and posttreatment prior to distribution. RO processes can produce water with TDS in the range 10–500 mg/L.

1.2.2.2.2 Nanofiltration

Nanofiltration (NF) is capable of removing many relatively larger organic compounds above about 300 Da and rejecting many divalent salts; monovalent ion removal can be in the range of 50%–90%. NF is applied in water softening, food, and pharmaceutical applications. Nanofiltration operates at lower pressure than RO

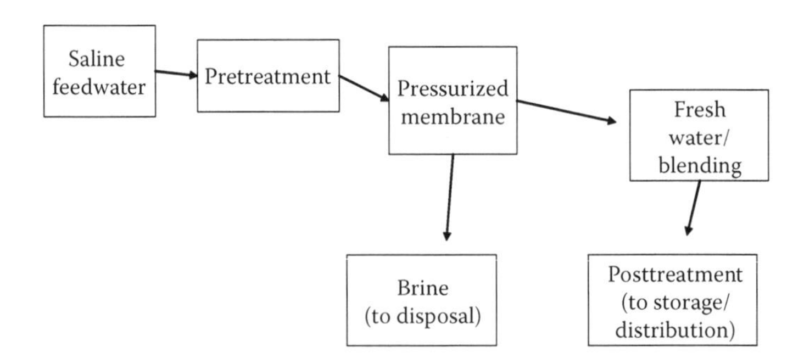

FIGURE 1.2 Reverse osmosis desalination process outline.

systems (e.g., ~340–3100 kPa). Systems may include several stages of polymeric membranes.*

1.2.2.2.3 Electrodialysis

In electrodialysis (ED)-based treatment systems, a direct current is passed through the water, which drives the ions (not the water) through membranes to electrodes of opposite charge. In electrodialysis reversal (EDR) systems, the polarity of the electrodes is reversed periodically during the treatment process. Ion transfer (permselective) anion and cation membranes separate the ions in the source water. Electrodialysis (ED and EDR) processes utilize selective membranes that contain cation and anion exchange groups. Under a direct current electric field, cations and anions migrate to the respective electrodes so that ion-rich and ion-depleted streams form in alternating spaces between membranes. Reversal of electric fields reduces scaling, organic fouling, and biological fouling, and flushes the membranes. Pretreatment is required to control scale and extend membrane life and to prevent migration of nonionized substances such as bacteria, organics, and silica.

1.2.2.2.4 Forward Osmosis, Pulsed Electrodialysis, and Captive Deionization

Forward osmosis, pulsed electrodialysis, and captive deionization are experimental approaches being studied at present. These experimental desalination technologies are in the early stages of development, and there are no practical applications of these experimental approaches in full-scale desalination plants at this time.

In forward osmosis, ammonia and carbon dioxide are added to fresh water on the opposite side of the membrane from the saline water to increase the ionic ammonium carbonate concentration so that water from the salt solution naturally migrates through the membrane to the ammonium carbonate "draw" solution without external pressure. The diluted "draw" solution is then heated to drive off the ammonia and carbon dioxide, which are captured and reused (Patel-Predd 2006). Potential advantages compared with RO include the fact that no external pressure is required, its high recovery efficiency, and lower energy costs. Additional research is required to determine its viability.

Pulsed electrodialysis could potentially eliminate the need for polarity and product/brine flow reversals of EDR and employs high direct current frequency to the ED stack, which reduces power use and increases ED membrane resistance to fouling.

Captive deionization draws ions to charged electrodes during a charging period and then releases them as brine (concentrate) during a discharge period. This technology does not incorporate a separate brine compartment, as with ED systems. Current voltage is kept below the threshold that would trigger redox reactions.

1.2.3 Pretreatment

Prefiltration, either by conventional treatment or by membranes, is essential to protect the RO membrane, reduce particulates and thus membrane fouling, to extend run

* Applied Membranes, Inc. (http://www.appliedmembranes.com); Dunlop Design Engineering Limited (http://www.dunlopdesign.com/).

life, and to improve the efficiency of the operation. Feedwater is treated to protect the membranes by removing some contaminants and controlling microbial growth on the membrane and to facilitate membrane operation. Suspended solids are removed by filtration, pH adjustments (lowering) are made to protect the membrane and control precipitation of salts, and antiscaling inhibitors are added to control calcium carbonates and sulfates. Iron, manganese, and some organics cause fouling of membranes. A disinfectant is added to control biofouling of the membrane. Disinfection can involve chlorine species, ozone or UV light, and other agents. Marine organisms, algae, and bacteria must be eliminated, and, if chlorine is used, it should be neutralized prior to contact with the membrane.

1.2.4 Posttreatment

Product water must be treated to stabilize it and make it compatible with the distribution system. Adjustment of pH to approximately 8 is required. Carbonation or other chemicals such as lime may be applied, and blending with some source water may be done to increase alkalinity and TDS and stabilize the water. Addition of corrosion inhibitors such as polyphosphates may be necessary. Postdisinfection is also necessary to control regrowth microorganisms during distribution, as well as to control pathogens from the blending process. Degasification may also be necessary. Many systems blend back a portion of the source water with the desalinated water for mineralization. With seawater, this is usually limited to about 1% due to taste contributed by sodium salts. Both blending with source water and treatment with lime or limestone also reconstitute some of the beneficial minerals. Some systems have utilized ED of seawater to generate hypochlorite *in situ* for disinfection and then blend it back to the desalinated water at about 1%. This practice generates very large amounts of bromate and organohalogen (halogenated) DBPs, and even the 1% blended desalinated water can far exceed the WHO provisional drinking water guideline for bromate of 10 μg/L (WHO 2008). It is being phased out for that reason.

1.3 Potentially Beneficial Chemicals

Water components can supplement dietary intake of trace micronutrients and macronutrients or contribute undesirable contaminants. The line between health and illness in a population is not a single bright line, but rather a complex matter of optimal intake versus adequate intake versus intake that is insufficient to maintain good health versus a toxic intake that will lead to frank illness in some higher-risk segments of the population. Some parts of the population, such as young children, pregnant women, the aged, the infirm, and the immunocompromised, can be more sensitive than the typical healthy adult to dietary components.

Some of the chemicals of beneficial interest in drinking water include calcium, magnesium, sodium, chloride, selenium, potassium, boron, iodide, fluoride, chromium, and manganese. Seawater is rich in ions such as calcium, magnesium, sodium, chloride, and iodine, but low in other essential ions, such as zinc, copper, chromium, and manganese. Desalination processes significantly reduce all of the ions in drinking water (compared with the original drinking water source), so that people who consume desalinated water may be consistently receiving smaller amounts of some nutrients relative to people who consume water from more traditional sources, and they are thus disadvantaged if their diets are not sufficient. As desalinated water can be stabilized by the addition of lime, for example, or sometimes by blending, some of these ions will be automatically replenished in that process (WHO 2005b, 2006).

Cardiovascular disease (CVD), osteoporosis, and prediabetic metabolic syndrome have been associated with dietary mineral deficiencies (WHO 2003, 2006). Based on the reviews of the several analytical epidemiological studies of drinking water conducted in Taiwan and Sweden and controlled human consumption studies in the United States, WHO expert consultations have concluded that the biological plausibility of a beneficial CVD effect is reinforced for persons with marginal magnesium consumption who are at higher risk of ischemic (sudden death) CVD. WHO is encouraging the conduct of robust before-and-after epidemiological studies, especially of sudden death rates due to heart disease among water consumers where the water hardness and especially the magnesium composition are being changed (Cotruvo and Bartram 2009). Before-and-after studies of ischemic CVD mortality can be conducted to determine if any change of incidence occurs that could be correlated with water quality changes. More comprehensive studies of total dietary magnesium and other minerals, including calcium, potassium, and sodium, relative to disease outcomes (e.g., CVD, osteoporosis, and diabetes) should be conducted where feasible.

Sodium can be present in desalinated water, depending on the efficiency of salts removal and the posttreatment blending, which could involve nondesalinated seawater. Typical total daily dietary intake of sodium can be in the range of 2,000–10,000 mg and more and is a function of personal taste and cultural factors. Water is usually not a significant contributor to total daily sodium intake except for persons under a physician's care who are required to be on highly restricted diets of less than 400 mg sodium/day. Although most people consume excess sodium and many persons are salt sensitive, some sodium in the water may be beneficial, especially for physically active people in warm climates.

The mineral composition of irrigation water is an issue in some soils. Magnesium, calcium, sulfate, and other ions are essential for plant growth. Excess boron has herbicidal activity, with sensitive crops in very low rainfall areas (Yermiyahu et al. 2007). Low-cost technologies are being developed for remineralization, including one that recycles magnesium from the reject brine water (Birnhack and Lahav 2007).

1.4 CONTAMINATION ISSUES

1.4.1 Source Contamination

Proper source selection and source protection are the best ways to avoid contamination of finished water by certain organics, surface runoff, ship discharges, and chemical and sanitary waste outfalls near the intake to the desalination plant. When contamination occurs, pretreatments may be necessary; these can involve enhanced disinfection and an adsorption process using granular activated carbon or, more frequently, powdered activated carbon for intermittent contamination. Of course, contaminants in blending waters will be transported to the finished water; thus, appropriate pretreatment of blending water may also be required.

1.4.2 Petroleum and Petroleum Products

The molecular weight cutoff for RO membrane performance is typically in the range of 100–300 Da for organic chemicals. Thus, RO membranes can efficiently remove larger molecules; however, small molecules are less efficiently removed, and some may pass through the membrane. Distillation processes can theoretically separate any substance by fractionation based on boiling point differences. However, distillation for desalination is not designed to be a fractionating system; thus, substances with boiling points lower than water's would be carried over in the vapors and should be vented out.

1.4.3 Disinfection and Microbial Control in Drinking Water

Similar to fresh waters, seawater and brine waters can contain pathogenic microorganisms, including bacteria, protozoa, and viruses. Disinfection can be applied at several points during the treatment process. The question is: What is the adequate level of disinfection to protect public health from exposure to pathogenic microbes, and are there any unique risks that may be associated with desalination practices?

During pretreatment, a disinfectant, often chlorine, will be added to reduce biofouling and protect the membrane from degradation. Membranes also have the capacity to remove microorganisms by preventing their passage to the finished water. So long as the appropriate membranes are intact, virtually complete removals of microorganisms can occur; however, some bacteria can grow through the membrane.

UF membranes, which have pores about 0.001–0.1 µm in size, have been demonstrated to achieve significant reductions of viruses and protozoa (R. McLaughlin, personal communication, 1998). Better performance would be expected from RO membranes. Several challenge tests employing *Giardia lamblia* cysts, *Cryptosporidium* oocysts, and MS2 bacteriophage with a UF membrane with a nominal pore size of 0.035 µm and an absolute pore size of 0.1 µm have demonstrated very effective removals. *Giardia* cysts can vary from 14 µm in length and from 5 to 10 µm in width; *Cryptosporidium* oocysts range in size from about 4 to 6 µm. Intact UF membranes (0.1 µm nominal) should completely remove the cysts. MS2 bacteriophage size is approximately 0.027 µm, which is smaller than the pore size of the UF membrane. However, substantial removal can be achieved, probably due to adsorption of the virus on suspended particles, adsorption on the membrane, or secondary filtration due to fouling of the membrane surface.

Permeation was observed in several cases in a bench-scale study (Adham et al. 1998) that evaluated rejection of MS2 coliphage (0.025 µm, icosahedral) by several commercial RO membranes with a nominal pore size cutoff of 0.001 µm. Although these were bench-scale simulations in particle-free water, they demonstrated that even similar RO membranes can have very different performance, and quality control procedures are required in their manufacture to ensure consistent performance for very small organisms such as viruses.

Distillation at high temperatures close to the normal boiling point of water would likely eliminate all pathogens. However, reduced pressures are used in some desalination systems to reduce the boiling point and reduce energy demands. Temperatures as low as 50°C–60°C may be reached. Several pathogenic organisms, including many protozoa, are denatured or killed in a few seconds to minutes at milk pasteurization temperatures in the 63°C (30 min) to 72°C (16 s) range, but spores and some viruses require higher temperatures and longer times.

Microbial growth during storage and distribution may be particular concerns when water is stored and distributed in very warm climates. Most regrowth microorganisms are not frank pathogens, but microorganisms such as *Legionella* that grow in plumbing systems at warm temperatures are a particular health concern and have caused numerous disease outbreaks, including deaths in hospitals and other buildings.

1.5 DISINFECTION BY-PRODUCTS

Significant amounts of DBPs can be formed in the pretreatments that are applied in both membrane and distillation processes. The desalination process must then be relied upon to remove them, along with the other contaminants that are present. Small solvent molecules such as trihalomethanes (THMs) will challenge the membranes; because many of the DBPs are volatile, they will also require venting during distillation processes. As desalinated waters are lower in TOC than many natural waters, it would be expected that the postdesalination disinfectant demand and also DBP formation would be relatively low; this has been indicated in some studies of THM production that have been reported (Al-Rabeh 2005). However, this could be significantly affected by the type of blend water that is used posttreatment to stabilize the water, particularly if seawater or other high-bromide water is used for blending. One of the factors to consider would be the amount of brominated organic by-products that could be formed if bromide is reintroduced to the finished waters, as well as subsequent bromate formation. As the TOC found in seawater could be different from the TOC in fresh waters, and as the pretreatment conditions are also different, it is probable that there would be differences in the chemistry of the by-product formation reactions that could lead to different amounts and types of by-products or different distributions of by-products.

1.6 WASTE AND CONCENTRATE MANAGEMENT

Wastes from desalination plants include concentrated brines, backwash liquids containing scale and corrosion salts and antifouling chemicals, and pretreatment chemicals in filter waste sludges. Depending on the location and other circumstances, including access to the ocean and sensitive aquifers, wastes could be discharged directly to the sea, mixed with other waste streams before discharge, discharged to sewers or treated at a sewage treatment plant, placed in lined lagoons, or dried and disposed in landfills. Concentrate disposal is one of the most challenging issues with respect to desalination processes. Recovery of important minerals from concentrates is possible and may be economically viable in some cases, because it also reduces waste disposal costs.

1.7 ENERGY CONSUMPTION

Desalination plants require significant amounts of electricity and heat, depending on the process, temperature, and source water quality. For example, it has been estimated that one thermal desalination plant producing about 26,500 m^3/day could require about 50 million kilowatt-hours per year, which would be similar to the energy demands of an oil refinery or a small steel mill. For this reason, cogeneration facilities provide significant opportunities for efficiencies. There is an obvious synergy between desalination and energy plants. Energy production plants require large water intakes for cooling purposes, they produce substantial amounts of waste heat that is usable in the desalination facility, and the spent water disposal system may also be shared. The International Atomic Energy Agency has studied the role of nuclear power plants as cogeneration facilities.

1.8 ENVIRONMENTAL IMPACTS

Installation and operation of a desalination facility will have the potential for adverse impacts on air quality, water/sea environment and groundwater, and possibly other aspects. All of these impacts must be considered, and their acceptability and mitigation requirements would usually be a matter of national and local regulation and policies. Studies to examine these effects would usually be conducted at each candidate site, and postinstallation monitoring programs should be instituted. A partial listing of issues follows:

- *Construction*: Coastal zone and seafloor ecology, bird and mammal habitat, erosion, nonpoint source pollution
- *Energy*: Fuel source and fuel transportation, cooling water discharges, air emissions from electrical power generation and fuel combustion
- *Air quality*: Energy production related
- *Marine environment*: Constituents in waste discharges, thermal effects, feedwater intake process, effects of biocides in discharge water, toxic metals, oxygen levels, turbidity, salinity, mixing zones, commercial fishing impacts, recreation, and many others
- *Groundwater*: Seepage from unlined drying lagoons causing increased salinity and possibly toxic metal deposition

REFERENCES

Adham, S., R.S. Trussell, P.F. Gagliardo et al. 1998. Rejection of MS2 virus by RO membranes. *JAWWA*, 90(9): 130–9.

Al Mutaz, I.S. 2000. Water desalination in the Arabian Gulf region. In *Water management, purification and conservation in arid climates. Vol. 2. Water purification*, ed. M.F.A. Goosen and W.H. Shayya, 245–65. Lancaster: Technomic Publishing Company.

Al-Rabeh, A. 2005. *Saudi Arabia report on desalination water quality data.* Engineering Department, Saline Water Conversion Corporation, Al-Khobar, Saudi Arabia.

Birnhack, L., and O. Lahav. 2007. A new post-treatment process for attaining Ca^{2+}, Mg^{2+}, SO_4^{2-} and alkalinity criteria in desalinated water. *Water Res.*, 41(17): 3989–97.

Cotruvo, J.A. 2005. Water desalination processes and associated health issues. *Water Conditioning & Purification Magazine* 47, no. 1: 13–7 (January). http://www.wcponline.com/pdf/0105%20Desalination.pdf.

Cotruvo, J.A. 2006. Health aspects of calcium and magnesium in drinking water. *Water Conditioning and Purification Magazine* 48, no. 6: 40–4 (June). http://www.wcponline.com/pdf/Cotruvo.pdf.

Cotruvo, J.A., and H. Abouzaid. 2007. New World Health Organization guidance for desalination: Desalinated water quality health and environmental impact. Paper presented at the Environment 2007 International Conference on Integrated Sustainable Energy Resources in Arid Regions, January 28–February 1, 2007, Abu Dhabi.

Cotruvo, J., and J. Bartram (eds.). 2009. *Calcium and magnesium in drinking-water: Public health significance*. Geneva: World Health Organization. http://www.who.int/water_sanitation_health/publications/publication_9789241563550/en/index.html.

EMS. 2006. Report that Arabs must spend $100 billion on water by 2016. Reported by Reuters UK, July 24, 2006. Dubai: Energy Management Services, Int. http://watersecretsblog.com/archives/2006/07/index.html.

GWI and IDA. 2006. *IDA desalination plants inventory*. International Desalination Association DesalData presented by Global Water Intelligence and Water Desalination Report. http://desaldata.com/.

Patel-Predd, P. 2006. Water desalination takes a step forward. *Environ. Sci. Technol.* 40(11): 3454–5.

UNEP. 2008. *Resource and guidance manual for environmental impacts assessments*. Manama: United Nations Environment Programme, Regional Office for West Asia; Cairo: World Health Organization, Regional Office for the Eastern Mediterranean. http://www.unep.org.bh/Newsroom/pdf/EIA-guidance-final.pdf.

USBR. 1976. Solicitation No. DS-7186. Denver: U.S. Department of the Interior, Bureau of Reclamation, 33.

WDR. 2006. *Water Desalination Report* 42, no. 35 (September). http://waterdesalreport.com/archive/42/35.

WDR. 2008. *Water Desalination Report* 44, no. 33 (September). http://waterdesalreport.com/archive/44/33.

WHO. 2003. *Nutrient minerals in drinking water and the potential health consequences of long-term consumption of demineralized and remineralized and altered mineral content drinking waters.* Report of the WHO workshop, Rome, November 11–13, 2003. Geneva: World Health Organization (WHO/SDE/WSH/04.01).

WHO. 2005a. *Water safety plans. Managing drinking-water from catchment to consumer.* Geneva, World Health Organization (WHO/SDE/WSH/05.06).

WHO. 2005b. *Nutrients in drinking water.* Geneva: World Health Organization. http://www.who.int/water_sanitation_health/dwq/nutrientsindw/en.

WHO. 2006. *WHO meeting of experts on the possible protective effect of hard water against cardiovascular disease.* Washington, D.C., April 27–28. Geneva: World Health Organization (WHO/SDE/WSH/06.06). http://www.who.int/water_sanitation_health/gdwqrevision/cardiofullreport.pdf.

WHO. 2008. *Guidelines for drinking-water quality.* 3rd ed. incorporating first and second addenda. Geneva: World Health Organization. http://www.who.int/water_sanitation_health/dwq/gdwq3rev/en/index.html.

Yermiyahu, U., A. Tal, A. Ben-Gal et al. 2007. Rethinking desalinated water quality and agriculture. *Science* 318: 920–1.

2 Desalination Process Technology

Nikolay Voutchkov, Corrado Sommariva, Tom Pankratz, and John Tonner

CONTENTS

2.1 GENERAL DESCRIPTION

This chapter analyzes the major aspects of desalination technology engineering and chemistry. It also identifies process-related mechanisms that may bring about departures from a desired drinking water quality goal. Environmental considerations are taken into account in the decisions for selection of the treatment processes and technologies and concentrate management. The chapter also highlights system designs and operational conditions that may bring about contamination or degradation of the desalinated product water, as well as design and operational practices that are normally adopted in order to avoid such departures. It could be considered as general guidance for process selection, design, and operation of desalination systems.

The chapter has been structured to address both major installations that produce large quantities of product water (i.e., plants with a production capacity of over 40,000 m^3/day) as well as small installations (e.g., package plants, shipboard systems, and point-of-use facilities) that are typically installed in remote areas in order to supplement existing water systems and where practical reasons often prevent the designers from adopting solutions and procedures that are widely used in large-scale installations.

2.1.1 Desalination Processes and Water Quality Issues

Water produced by desalination methods has the potential for contamination from source water and from the use of various chemicals added at the pretreatment, desalination, and posttreatment stages. Natural water resources are more likely to be impacted by microbiological contamination when they receive wastewater discharges and surface runoff.

Regulatory frameworks developed for water produced by desalination take account of these differences; in addition, they consider the type of desalination process employed, using either thermal or membrane technology. Testing processes are discussed in Chapter 5; they assume that monitoring is designed for operational control as well as to meet regulatory quality requirements. Monitoring is not an end in itself; rather, it should be designed to provide useful information to confirm that the process was properly designed and built and is being properly operated to prevent contamination from reaching consumers.

2.1.2 Water Safety Plans

The safety and performance of a system for providing drinking water depend on the design, management, and operation of the three principal components: source, treatment, and distribution. If contamination has occurred and is not controlled before it reaches the consumers' taps, illness or even death is possible. Thus, the entire system must be designed to anticipate and cope with all of the problems that could occur, and proper performance of the entire system must be ensured at all times.

The most effective way to consistently ensure the safety of a drinking water supply is through the use of a comprehensive risk assessment and risk management approach that encompasses all of the steps in the water supply train from the source water catchments to the consumer. This concept is fully applicable to desalination systems. The World Health Organization (WHO) has developed the systematized water safety plan (WSP) approach based on the understanding of water system function derived from the worldwide history of successful practices for managing drinking water quality (WHO 2008; see also Chapter 5). The WSP concept draws upon principles and concepts of prevention, multiple barriers, and quality management systems, such as the hazard assessment and critical control point approach used in the food industry. Desalination treatment processes are usually more comprehensive than standard water technologies, so they are particularly suited to WSP applications.

A WSP has three key components guided by health-based targets (drinking water standards, guidelines, and codes) and overseen through surveillance of every significant aspect of the drinking water system. The three components are

System assessment to determine whether the system as a whole (from source to consumer) can consistently deliver water that meets health-based targets; this includes assessment of design criteria for new systems, as well as modifications.

Measures to monitor and control identified risks (and deficiencies) and ensure that health-based targets are met; for each control measure, appropriate operational monitoring should be defined and instituted that will rapidly detect deviations.

Management plans describing actions to be taken during normal operations or incident conditions and documenting the system assessment (including system upgrades and improvements), monitoring, communication plans, and supporting programs.

The primary objectives of a WSP are the minimization of contamination of source waters, reduction or removal of contamination through appropriate treatment processes, and prevention of contamination during processing, distribution, and storage. These objectives are equally applicable and can be tailored to large piped supplies, small community supplies, large facilities (hotels and hospitals), and even household systems. The objectives are met through the interpretation and detailed implementation of the key phrases "hazard assessment" and "critical control points" in a systematic and documented planned methodology for the entire life of the system. A progression of the key steps in developing a WSP is as follows:

- Assemble and train the team to prepare the WSP.
- Document and describe the system.
- Undertake a detailed hazard assessment and risk characterization to identify and understand how hazards can enter the system.
- Assess the existing system (including a description of the system and a flow diagram).
- Identify control measures: the specific means by which specific hazards may be controlled.
- Define monitoring of the control measures: the limits that define acceptable performance and how these are monitored.
- Establish procedures to verify that the WSP is functioning effectively and will meet the health-based targets.
- Develop supporting programs (e.g., training, hygiene practices, standard operating procedures, upgrades and improvements, research and development).
- Develop management procedures, including corrective actions for normal and incident conditions.
- Establish documentation and communication procedures.

These key steps in the WSP operate in a continuous and cyclical mode by returning to the documenting and system description step and repeating the process routinely. A detailed treatment of this WSP concept can be found in the WHO *Guidelines for Drinking-Water Quality* (WHO 2008).

2.2 STRUCTURE OF CHAPTER

The description of desalination treatment technologies presented herein follows the treatment process sequence of a typical desalination plant (see Figure 2.1). For each

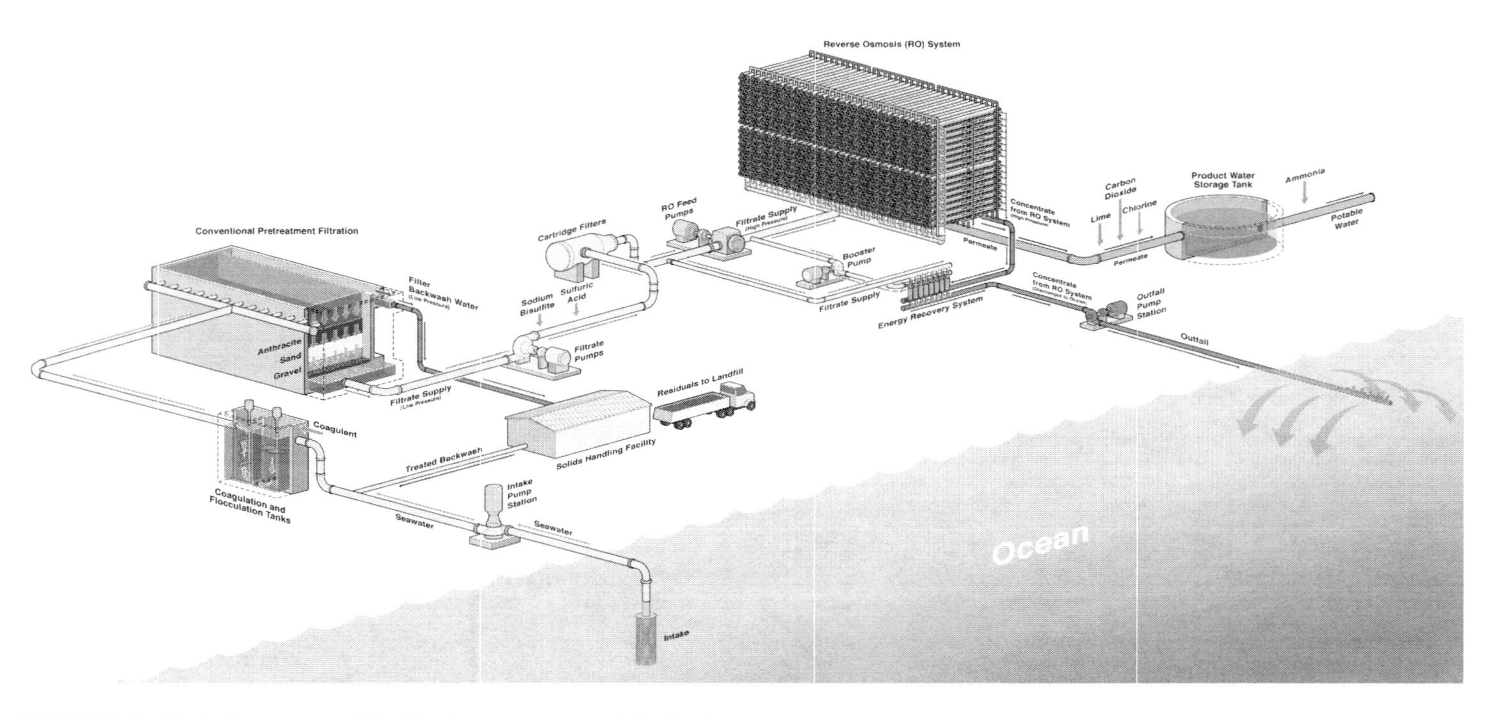

FIGURE 2.1 Typical sequence of desalination treatment and distribution processes.

treatment process step, there is a description of the main treatment technologies that are widely used, followed by a highlighting of the issues and considerations for that step.

2.3 SOURCE WATER INTAKE FACILITIES

2.3.1 General Description

Desalination facilities require an intake system capable of providing a reliable quantity of source water (raw feedwater) of a reasonably consistent quality and with a minimum ecological impact. As the first step in the pretreatment process, the type of intake used would affect a range of source water quality parameters and would impact the performance of downstream treatment facilities. Intake designs are highly site specific, possibly more so than any other aspect of the desalination facility. The design, modeling, monitoring, and permitting activities that surround them may represent as much as 10%–30% of the capital cost of the entire facility. Not only would a good intake design protect downstream equipment and reduce environmental impact on aquatic life, but it would enhance process performance and reduce pretreatment system capital and operating costs as well.

Two general types of intake facilities are used to obtain source water for desalination plants: subsurface intakes (wells, infiltration galleries, etc.) and open intakes. Seawater intake wells are either vertical or horizontal source water collectors, which are typically located in close proximity to the sea. In the case of aquifers of high porosity and transmissivity, which easily facilitate underground seawater transport such as the limestone formations of many Caribbean islands and Malta, seawater of high quality and large quantity may be collected using intake wells located inland rather than at the shore. This allows the desalination plant to be located closer to the main users rather than at the shore, reducing the distance for seawater collection and thus the costs of conveyance. Brackish water treatment plants usually use wells for source water collection, as the source water is typically located above inland brackish aquifers.

Intake wells are relatively simple to build, and the seawater or brackish water they collect is pretreated via slow filtration through the subsurface sand or seabed formations in the area of source water extraction. Vertical intake wells are usually less costly than horizontal wells; however, their productivity is relatively small, and the use of vertical wells for large plants is therefore less favorable (Figure 2.2).

Horizontal subsurface intakes are more suitable for larger seawater desalination plants and are applied in two configurations: radial Ranney-type collector wells (Figure 2.3) and horizontal directionally drilled (HDD) collectors. Vertical wells and radial collector wells are used to tap into the onshore coastal aquifer or inland brackish water aquifer, whereas HDD collectors are typically extended offshore under the seabed for direct harvesting of seawater. The HDD collector wells consist of relatively shallow blank well casing with one or more horizontal perforated screens bored under an angle (typically inclined at 15°–20°) and extending from the surface entry point underground past the mean tide line at a minimum depth below the seafloor of 5–10 m.

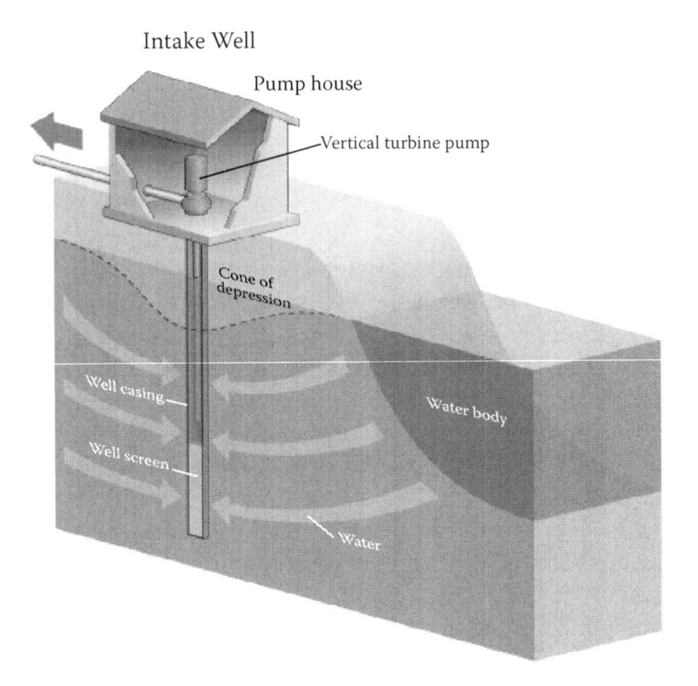

FIGURE 2.2 Vertical intake well.

Open ocean intakes are suitable for all sizes of seawater desalination plants but are typically more economical for plants with a production capacity higher than 20,000 m^3/day. Open intakes for large seawater desalination plants are often complex structures, including intake piping that typically extends several hundred to several thousand meters into the ocean. Source water collected through open intakes usually requires pretreatment prior to reverse osmosis desalination. The cost and time for construction of a new open ocean intake could be significant and could reach 10%–20% of the overall desalination plant construction cost. Open ocean intakes would result in some entrainment of aquatic organisms compared with beach wells, because they take raw seawater directly from the ocean rather than source water prefiltered through the coastal sand formations. Subseabed horizontal intakes have the benefit of providing some filtration pretreatment while causing minimal entrainment of marine life and having a limited aesthetic impact on the shoreline (Peters and Pinto 2006).

Raw seawater collected using wells is usually of better quality in terms of solids, silt, oil and grease, natural organic contamination, and aquatic microorganisms, compared with open seawater intakes. Well intakes may also yield source water of lower salinity than open intakes. However, they have the potential for altering the flows of hydraulically connected freshwater aquifers and possibly accelerating seawater intrusion into these aquifers. Use of subsurface intakes for large desalination plants may be limited by a number of site-specific factors that should be taken into consideration when selecting the most suitable type of intake for a particular project (Watson et al. 2003; Voutchkov 2004a).

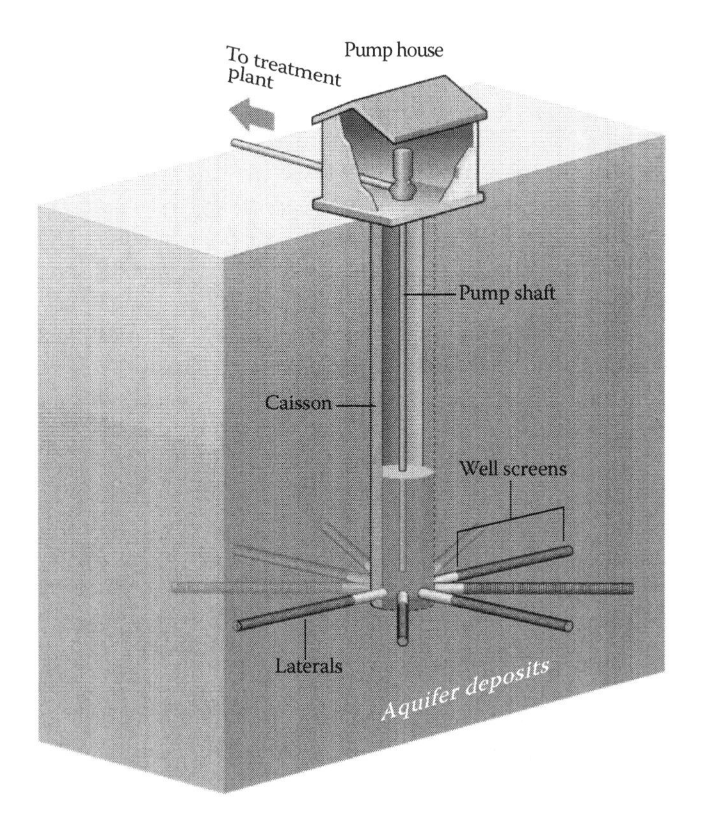

FIGURE 2.3 Horizontal intake well.

2.3.2 Source Water Issues and Considerations

2.3.2.1 Water Quality

A thorough raw water characterization at the proposed intake sites must include an evaluation of physical, microbial, and chemical characteristics, meteorological and oceanographic data, and aquatic biology. An appropriate intake design must also consider the potential effects of fouling, continuous or intermittent pollution, and navigation and must take necessary steps to mitigate these source water contamination, environmental, and operational risks. Seasonal variations should also be characterized and understood, preferably before the desalination plant is designed and built. Chapter 5 provides a detailed description of suggested water quality monitoring needs for desalination plants.

Similar to most process systems, desalination plants operate most efficiently and predictably when feedwater characteristics remain relatively constant and are not subject to rapid or dramatic water quality fluctuations. Therefore, the water quality review should consider both seasonal and diurnal fluctuations. The review should consider all constituents that may have an impact on plant operation and process performance, including water temperature, total dissolved solids (TDS), total suspended solids

(TSS), membrane scaling compounds (calcium, silica, magnesium, barium, etc.), and total organic carbon (TOC). Desalination intake water requirements and quality vary based on the desalination process employed. Feedwater volume requirements generally range from approximately 25% more than the production capacity of some brackish water reverse osmosis (BWRO) plants to two times the plant production capacity for seawater reverse osmosis (SWRO) systems. Thermal seawater desalination systems often require more than 10 times the distillate production, because they have both process and cooling water requirements. The necessary feedwater must be available whenever a plant is operating if the plant is to meet productivity goals.

In addition to the permeability, productivity, and safe yield of the source water aquifer, the key factor that determines the location and feasibility of the intake for a brackish water desalination plant is the raw water quality. Plant failure can be caused by large changes of intake water salinity and variations in or elevated concentrations of water quality contaminants such as silica, manganese, iron, radionuclides, or scaling compounds, which may create fouling or operational problems that increase treatment costs or limit the available options for concentrate disposal. Subsurface geological conditions determine to a great extent the quantity and quality of the raw water. Confined or semiconfined aquifers yield the most suitable source of water for brackish water desalination systems (Missimer 1999).

Whenever groundwater is pumped from an aquifer, there is always some modification of the natural flow in that aquifer. Some brackish water aquifers are density stratified; when water is pumped from the top portion of the aquifer, higher-salinity groundwater propagates upward, increasing source water salinity over time. Many brackish water aquifers are semiconfined, and they may have a common boundary with other aquifers of different water quality. When the production aquifer is pumped, a certain portion of the recharge volume to this source water aquifer may be supplied from the adjacent bounding aquifers, thus causing a change of source water quality from that of the original aquifer to the water quality of the bounding aquifers. These changes in source water quality over time may affect not only the intake water salinity but also the overall ionic makeup of the source water for brackish water desalination plants, which in turn may affect the system's allowable recovery and may also impact concentrate disposal permitting. Therefore, it is essential to conduct predesign hydrogeological investigations that include predictive modeling of the potential long-term changes in source water quality that could occur over the useful life of a subsurface intake system. Protection of freshwater aquifers is an essential consideration.

2.3.2.2 Impacts on Aquatic Life

Environmental impacts associated with concentrate discharge have historically been considered the greatest single ecological impediment in selecting the site for a desalination facility. However, aquatic life impingement and entrainment by the desalination plant intake are more difficult to identify and quantify and may also result in measurable environmental impacts.

Impingement occurs when aquatic organisms are trapped against intake screens by the velocity and force of flowing water. Entrainment occurs when smaller organisms pass through the intake screens and into the process equipment. The results of impingement and entrainment vary considerably with the volume and velocity of feedwater and the use of mitigation measures developed to minimize their impact. Impingement and entrainment of aquatic organisms are not environmental impacts unique to open intakes of seawater desalination plants. Conventional freshwater open intakes from surface water sources (i.e., rivers, lakes, estuaries) may also cause measurable impingement and entrainment. There are a number of surface and subsurface intake options that may be employed to mitigate these and other environmental impacts, and all options should be thoroughly examined.

2.3.2.3 Water Discoloration and Taste and Odor Issues Associated with Anaerobic Wells

Some seawater and brackish water wells draw their water supply from anaerobic aquifers that may contain hydrogen sulfide. In those instances, the feedwater intake and conveyance systems should remain pressurized to prevent the formation of elemental sulfur. After desalination, product water must be degasified to prevent taste and odor problems. Some brackish waters from anaerobic wells may contain large amounts of iron and manganese in reduced form. Exposure to oxygen may oxidize the iron and manganese salts and discolor the source and product water and the plant discharge. In this case, the source water should be treated with greensand filters to oxidize and remove iron and/or manganese salts under controlled conditions.

2.3.2.4 Biofouling

All natural water systems contain a wide range of microorganisms that can cause operational problems if not controlled. These organisms grow predominantly in slime-enclosed biofilms attached to surfaces. Biofilms may form very rapidly, restricting the flow of water through the reverse osmosis (RO) membranes. The formation of a biofilm of microorganisms on the surface of the desalination membranes of SWRO and BWRO plants and on the contact surfaces of thermal desalination plants to an extent that causes a measurable reduction of the production capacity of the desalination system is typically referred to as *biofouling*. Although most aquatic organisms causing biofouling are not pathogens, their excessive growth could have a negative effect on the desalination plant's overall performance and efficiency. Once a biofilm establishes itself and fouls a membrane, it may be extremely difficult, or even impossible, to remove it. Desalination plants with open intakes typically incorporate facilities for biofouling control that include the use of chlorine or other oxidants or biocides to control excessive biogrowth. Thermal desalination plants usually apply continuous chlorination, whereas most membrane systems practice intermittent or shock chlorination.

Aquatic organisms, including mussels, barnacles, clams, and mollusks, may grow in intake channels, pipes, and equipment, causing operational problems. Open intake systems are usually equipped with provisions for hindering biogrowth and for periodic removal of aquatic organisms from the plant intake facilities in order to maintain reliable and consistent plant performance.

2.3.2.5 Collocation of Desalination Plants and Power Generation Facilities

Because power generation plants require large volumes of cooling water to condense the power cycle steam, desalination facilities often consider collocation with a power plant with which they can share a common intake. The avoided cost of constructing and permitting a new intake may reduce the capital cost of a large desalination facility by 10%–20%. A more detailed description of the collocation configuration is provided in Section 2.8. However, impingement and entrainment considerations with once-through power plant cooling systems may affect the permitting of collocated desalination facilities. Power plants and desalination facilities proposing to share seawater intakes must identify how the operation of both the power plant and the desalination plant will be coordinated to minimize impingement and entrainment.

Recent studies indicate that the incremental entrainment effect of desalination plant intakes collocated with once-through power plants is minimal. For example, the entrainment study completed for the 200,000 m^3/day Huntington Beach Seawater Desalination Project in California, and included in the environmental impact assessment (EIA) report for this project (City of Huntington Beach 2005), determined that the collocation of this plant with the existing AES Power Generation Station allows the reduction of the additional entrainment effect attributed to the desalination plant intake to less than 0.5%. A similar EIA study in Carlsbad, California (City of Carlsbad 2005), showed that the maximum entrainment potential of the proposed 200,000 m^3/day seawater desalination plant is reduced to less than 1% as a result of the collocation of this plant's intake with the Encina Power Generation Station's discharge. Because those desalination plants use warm cooling water collected from the power plant discharge and do not have separate new open ocean intakes and screening facilities, they would not cause an incremental impingement of aquatic organisms.

2.4 PRETREATMENT PROCESSES

2.4.1 General Description

Screening of the intake water is the first step of the treatment process. The primary function of pretreatment is to ensure that turbidity/suspended solids and the quantity of organic and inorganic foulants are within the acceptable range for the desalination process equipment. Secondary functions may include the removal of other unwanted constituents that may be present (continuously or intermittently), such as hydrocarbons or algae.

The pretreatment process improves the quality of the raw feedwater to ensure consistent performance and the desired output volume of the desalination process. Almost all desalination processes require pretreatment of some kind. The level and type of pretreatment required depend on the source and quality of the feedwater and the chosen desalination technology. Pretreatment can constitute a very significant portion of the overall plant infrastructure for source water of poor quality. The potential influences on public health and the environment from the pretreatment process operations are associated with the chemical conditioning (addition of biocides, coagulants, flocculants, antiscalants, etc.) of the source water prior to pretreatment and with the disposal of the residuals formed during the pretreatment process. Pretreatment, when required, normally involves a form of filtration and other physicochemical processes whose primary purpose is to remove the suspended solids (particles, silt, organics, algae, etc.) and oil and grease contained in the source water when membrane desalination is used for salt separation. For thermal desalination processes, pretreatment protects downstream piping and equipment from corrosion and from formation of excessive scale of hard deposits on their surface (scaling). Biofouling is most often mitigated using an oxidant, although nonoxidizing biocides are also utilized. Potential concerns associated with pretreatment are typically associated with the by-products formed during the chemical conditioning process and their potential propagation into the finished fresh water.

2.4.2 Pretreatment for Thermal Desalination Plants

For thermal desalination facilities, the pretreatment process must address

- Scaling of the heat exchanger surfaces, primarily from calcium and magnesium salts (acid-treated plants)
- Corrosion of the plant components, primarily from dissolved gases
- Physical erosion by suspended solids
- Effects of other constituents, such as oil, growth of aquatic organisms, and heavy metals

Thermal desalination systems are quite robust and normally do not include any physical treatment other than what is provided by the intake (i.e., no additional filters or screens). Chemical conditioning is utilized in thermal desalination in two treatment streams: the cooling water (which is the larger flow and is generally returned to the feed source) and the makeup water (used within the desalination process). Cooling water is normally treated to control fouling using an oxidizing agent or biocide. The makeup water is continuously treated with scale inhibitors (usually a polymer blend) and may be intermittently dosed with an antifoam surfactant (typically during unusual feedwater conditions).

2.4.3 Chemicals Used in Thermal Desalination Processes

Table 2.1 profiles chemicals that are most frequently used in seawater pretreatment for thermal desalination. Dose rates are only indicative and are shown as concentrations of chemical in the relevant process stream.

2.4.4 Source Water Pretreatment for Membrane Desalination

Membrane desalination is used to desalinate water from many sources, including brackish surface water from rivers and lakes, brackish groundwaters, municipal and industrial wastewater, and seawater from open ocean intakes and beach wells. Because of the great variability of the water quality, depending on its source

TABLE 2.1
Chemicals Used in Thermal Desalination Processes

Chemical Type	Purpose of Use	Dose and Feed Location	Application
Scale inhibitor (usually phosphonates, polyphosphate, polymaleic, or polycarboxylic acids, or a blend of several of these)	Usually crystal modifiers that avoid precipitation and development of deposits (primarily calcium carbonate, magnesium hydroxide); blends may include dispersant properties to prevent crystals adhering to equipment	1–8 mg/L, MW	Used in all thermal desalination processes
Acid (usually sulfuric acid)	An alternative scale inhibitor; calcium carbonate and magnesium hydroxide scale formation is avoided by lowering the pH	~100 mg/L, MW	Used only in MSF desalination
Antifoam (polypropylene/ polyethylene oxide or similar surfactant)	Uncorrected foaming due to unusual feedwater conditions may overwhelm the process, indicated by high product TDS (carryover)	~0.1 mg/L, MW	Used intermittently in all thermal processes, but primarily MSF
Oxidizing agent: most often a form of chlorine; however, biocides may have some use, particularly for smaller systems	To control biofouling and aquatic organism growth in the intake and desalination equipment; continuous dosing of 0.5–2 mg/L active chlorine with intermittent shock dosing (site specific, but may be 3.7 mg/L for 30–120 min every 1–5 days)	~1.0 mg/L, CW	Used for large surface water and seawater intakes
Sodium bisulfite	Oxygen scavenger to remove traces of residual oxygen or chlorine in the brine recirculation	~0.5 mg/L, MW	Used only in MSF desalination systems and in intermittent mode

Note: CW = cooling water; MW = makeup water.
Source: Courtesy of N. Voutchkov.

(brackish water or seawater) and the type of intake (open or subsurface), simple generalizations about pretreatment requirements are not definitive.

For membrane desalination facilities, the pretreatment processes must address

- Membrane fouling and scaling from metal oxides, colloids, and inorganic salts
- Fouling or plugging by inorganic particles
- Biofouling by organic materials and microorganisms
- Chemical oxidation and halogenation by residual chlorine

- Chemical reduction of chlorine
- Effects of other constituents, such as oil, aquatic organisms, and heavy metals

Membrane desalination requires a higher degree of pretreatment than thermal desalination processes. Membrane separation technologies were developed for the removal of dissolved salts, but they also block the passage of filterable materials. Membranes are not designed to handle high loads of filterable solid materials, and the presence of suspended solids in the source water can reduce the quality and quantity of water produced or lead to shorter-than-anticipated membrane life and inferior membrane performance. Filterable solid materials are removed by the pretreatment process to achieve low content of suspended solids and silt in the water, which is measured by a cumulative parameter called the *silt density index* (SDI). SDI values of the source water are indicative of its membrane-fouling tendency and are calculated using a procedure that includes filtration of the water sample at constant pressure through a 0.45 μm filter. Generally, most membranes require feedwater with an SDI of less than 5 in order to maintain steady and predictable performance.

2.4.5 Chemicals Used for Pretreatment Prior to Membrane Desalination

Table 2.2 profiles some chemicals that are used for pretreatment of the source water prior to membrane desalination. Some chemicals are used continuously to optimize operations, whereas others are used intermittently for cleaning of the filtration media of the pretreatment system. Dose rates are only indicative and are shown as milligrams of chemical per liter of feedwater.

The chemicals listed in Table 2.2 are typically used for conventional granular media (anthracite and sand) pretreatment systems. Microfiltration (MF) or ultrafiltration (UF) membrane systems can also be used as a pretreatment to desalination. Although, typically, the use of MF and UF pretreatment systems does not require source water conditioning, these systems use a significant amount of chemicals for chemically enhanced backwash (CEB) and cleaning of the pretreatment membranes. Typically, CEB is practiced once or twice every day, whereas deep chemical cleaning of the MF/UF membranes is completed every 60–90 days. Table 2.3 summarizes the types of chemicals used for membrane pretreatment.

All chemicals utilized in membrane desalination applications should always have an appropriate certification and quality control to confirm their suitability for use in drinking water. Typically, specialized governmental, public, or private organizations certify products to those standards.

2.5 THERMAL DESALINATION PROCESSES

When saline solutions are boiled, the escaping vapor contains water-soluble gases and volatile organics (which are vented), whereas the salts and some organics remain in the unevaporated solution. The evaporation-based salt separation process yields water of very low salt content (usually below 5 mg TDS/L), but the latent heat required to evaporate the water is high. As a result, several process configurations

TABLE 2.2
Pretreatment Chemicals Used in Membrane Desalination Systems

Chemical Type	Purpose of Use	Dose	Application
Scale inhibitors (polyelectrolyte polymer blends)	Increase of solubility of sparingly soluble salts such as calcium and magnesium carbonates and sulfates; additional chemicals may be used to target specific species, such as silica	~2–5 mg/L	Primarily in brackish water desalination and water reclamation using RO and ED/EDR operating at high recoveries
Acid (usually sulfuric or hydrochloric acid)	Reduction of pH for inhibition of scaling and for improved coagulation	40–50 mg/L as required to reduce pH to ~6–7	Primarily in seawater RO applications; not used in all applications
Coagulant (usually ferric chloride or ferric sulfate)	Improvement of suspended solids removal	5–15 mg/L	Primarily in open intake seawater RO and surface water RO systems
Flocculant aid (usually cationic polymer)	Improvement of suspended solids removal	1–5 mg/L	Primarily in open intake seawater RO and surface water RO; may be used only intermittently when feed SDI is unusually high
Oxidizing agent: most often a form of chlorine; however, biocides have found some use, particularly in smaller systems	To control biofouling and aquatic organism growth in the intake and pretreatment facilities; chloramines may be used for pretreatment in reclamation systems, and their use should be avoided in seawater desalination systems	Site specific, but may be 3.7 mg/L for 30–120 min every 1–5 days	Used for large surface water and seawater intakes; small systems and those using wells, especially those in which source water is anaerobic, may not require oxidation
Reducing agent (usually a form of bisulfite); function of chlorine dosage	To eliminate oxidizing impacts on the RO membrane	Generally 2–4 times higher than oxidizing agent dose	In all membrane processes using polyamide RO membranes (less common cellulose acetate membranes have greater tolerance for oxidants)
Membrane preservation and sterilization	Offline membranes must be sterilized and preserved; sterilization may utilize hydrogen peroxide; in some cases, acetic acid is also used to create peracetic acid; preservation most commonly utilizes sodium bisulfite		

Source: Courtesy of N. Voutchkov.

TABLE 2.3
Chemicals Used for Cleaning Membrane Pretreatment Systems

Chemical Type	Purpose of Use	Dose	Application
Acid (usually citric, phosphoric, or hydrochloric acid)	Cleaning of solids and biological material from membrane filtration	Batch size is a function of process train size; frequency of batches is a function of number of process trains and site-specific conditions	MF/UF membrane CEB and periodic cleaning
Sodium hypochlorite	Cleaning biological material from membrane filtration	Batch size is a function of process train size; frequency of batches is a function of number of process trains and site-specific conditions	MF/UF membrane CEB
Phosphates (tripolyphosphate or similar)	Cleaning of membranes	Batch size is a function of process train size; frequency of batches is a function of number of process trains and site-specific conditions	Periodic MF/UF membrane cleaning
Ethylenediamine-tetraacetic acid (EDTA)	Cleaning of membranes	Batch size is a function of process train size; frequency of batches is a function of number of process trains and site-specific conditions	Periodic MF/UF membrane cleaning
Specialty cleaning chemicals	Unusual deposits on membrane surfaces may be removed, offline, using specific chemicals and treatments specified by the membrane manufacturers	This may be not only offline, but in some cases off-site	Periodic MF/UF membrane cleaning

Source: Courtesy of N. Voutchkov.

have been developed in an attempt to minimize energy consumption. The two most widely used thermal desalination processes are multistage flash distillation (MSF) and multiple effect distillation (MED). Both MSF and MED can be used for desalinating seawater and brackish water. However, the majority of the existing MSF and MED plants are seawater desalination facilities.

2.5.1 MSF Desalination

Until the early 1990s, MSF was the most commonly employed method of seawater desalination. In the MSF process, a stream of heated seawater flows through the bottom of a vessel containing up to 40 chambers or stages, each operating at a slightly lower pressure than the previous one (Figure 2.4).

The lower pressure causes the hot seawater to begin boiling immediately upon entering each new stage. The rapid, violent boiling action causes a portion of the seawater to instantly vaporize, or flash into steam. The flashing process cools the concentrated seawater and allows it to reach thermodynamic equilibrium with its new surroundings. Flashed vapor rises rapidly, passing through wire mesh demister pads to remove entrained *brine* droplets. The cleansed vapor then passes around the outside of the tube bundle carrying cool seawater, where it is condensed into pure distilled water. Distillate is collected in a trough, and it flows into the distillate section of the following stage. Because the stage is operating at a lower pressure, a small portion of the distillate flashes in a manner similar to the concentrated seawater. This flashing reduces the temperature of the distillate until it reaches the last stage of the process, where its temperature is usually >3°C–5°C higher than that of the seawater.

Oxygen, carbon dioxide, and other gases dissolved in seawater are released during the evaporation process. The vacuum in which the system operates may also draw air into the evaporation vessels (effects) through small leaks in the vessel walls. If these noncondensable gases (NCGs) are allowed to accumulate within the system, they may hinder heat transfer by a process known as *gas blanketing*. A venting system must be incorporated within the system to ensure that the NCGs are continually swept away from the heat transfer surfaces. The venting systems usually use a combination of steam jet ejectors and small condensers to completely evacuate the NCGs to the atmosphere.

MSF is considered a forced circulation process because pressure inhibits evaporation of the hot brine until it passes through an orifice into a chamber below its boiling temperature. Upon entering the lower-pressure chamber, a portion of the brine instantly flashes to vapor without the addition of heat. The vapor is naturally drawn toward the cooler tube surfaces, where it is condensed. The heat of condensation passes through the tubes, heating the seawater inside. The stage remains at a constant temperature and pressure and is essentially self-regulating. The seawater temperature increases progressively as it passes through the tubes, stage by stage, until it reaches the first—and hottest—stage. From stage 1, the preheated seawater enters a separate steam condenser, or brine or feedwater heater, where motive steam from a boiler or other source is used to heat the seawater to the top brine temperature (TBT), which is typically 90°C–115°C. The pressure within the feedwater heater ensures that no boiling or flashing occurs before the hot seawater is reintroduced to the shell side of the tubes in the first stage. At this point, the brine flow resembles a turbulent, rapidly flowing river that may be up to 20 m wide and 100 m long in the largest MSF systems. Most plants occupy a single level, but several have double levels, or decks, as a method to reduce the footprint.

Large-scale MSF systems are used for municipal water supply, often in dual-purpose power and water cogeneration facilities. Steam from back-pressure or condensing turbines typically provides the heat to drive the process. These systems are used in applications where there is a large demand for water coincident with a relatively large amount of (waste) heat. Large systems produce between 10,000 and 65,000 m^3 of water per day from each machine.

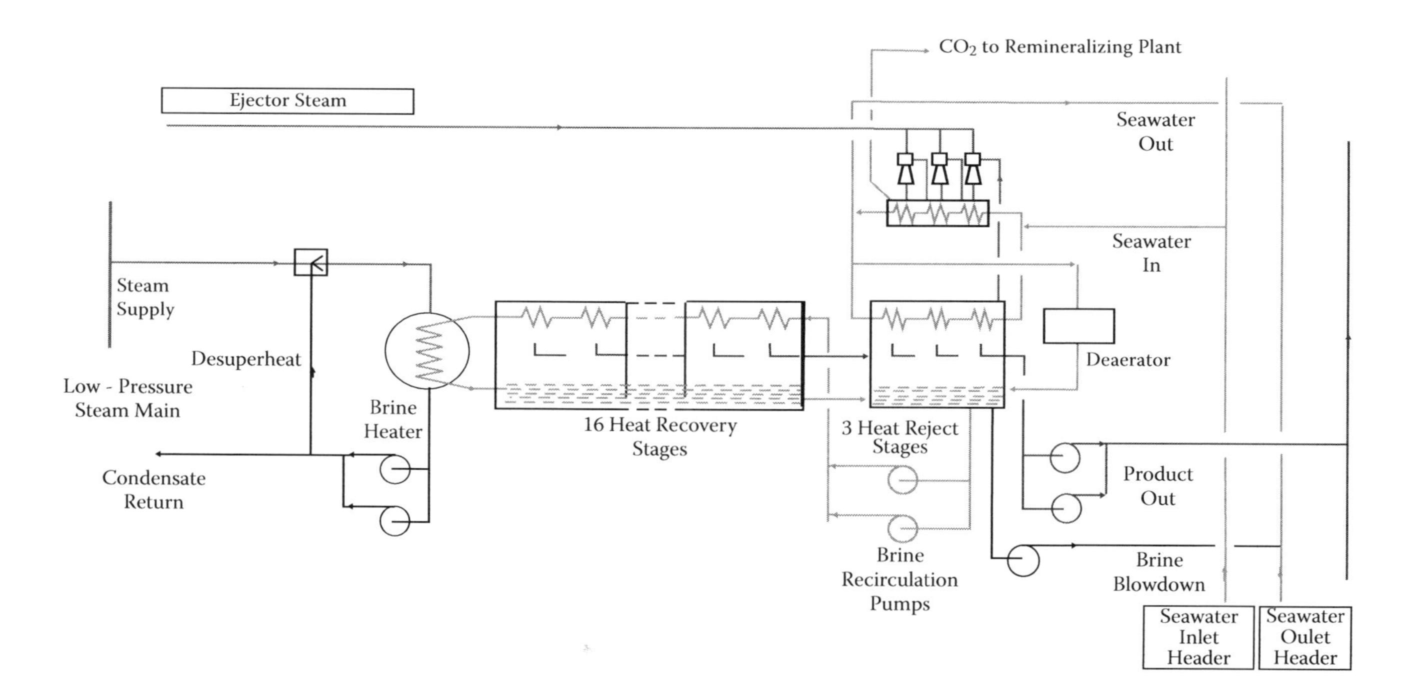

FIGURE 2.4 Schematic of a typical MSF thermal desalination system.

2.5.1.1 MSF Issues and Considerations

The cooling water returned to the sea is generally 8°C–12°C warmer than the ambient conditions. It is not deaerated and may contain small quantities of corrosion by-products or disinfection by-products (DBPs). MSF systems generally produce distilled water with a TDS content of between 2 and 50 mg/L. The distillate TDS comes from solids carryover or possible joint leakage or tube failure allowing bypassing of the separation process. The integrity of the tubes and joints can be checked and confirmed by hydrotesting; however, failures are readily identified by rapid increases in distillate TDS. NCGs are redissolved in the distillate, the primary impact of which is the reduction of pH to around 6.2 (due to carbon dioxide). The small number of facilities that use acid to dissolve metal salts for scale control will produce distilled water with a more neutral pH. There is also the potential to distill volatile materials that may be in the feedwater source, unless these are removed by the external pretreatment process or the deaeration step, which is an integral part of an MSF system.

The process concentrate, frequently referred to as brine, is returned to the source (usually the sea). Most commonly, the brine is blended with cooling water from the MSF process (as well as cooling water from other processes, if available). The brine contains scale inhibitor and antifoam chemicals and is generally considered deaerated (prior to blending with cooling water). The brine TDS concentration is generally 1.4–1.8 times higher than that of the raw seawater/makeup water. The brine is usually >3°C–5°C warmer than the ambient seawater (before blending with cooling water).

There is no direct contact between the heating steam system and the desalination process. Steam condensate does not mix with the seawater, brine, or distilled water. However, in some systems, there can be a vent connection between the steam system and the vapor space above the flashing brine, which can introduce approximately 0.5%–1.0% of the steam flow into the MSF process; vapor contaminants should be swept into the vacuum system. This vent connection provides useful heat recovery, but usually can be (at some reduction in efficiency) directed to the more isolated vacuum system if the steam is thought to contain contaminants. Condensate drains from the vacuum system could also contain dissolved or absorbed contaminants from seawater decomposition and/or from steam. These drains sometimes connect into the brine side of the process (for some minimal heat recovery prior to rejection from the process); however, they can also be directly rejected if required. Similarly, there should be consideration of possible contamination for special heat recovery systems (such as drain coolers) that, while not common, may still be utilized.

These systems can be manufactured from a variety of materials, but alloys of copper and nickel and various molybdenum-bearing grades of austenitic stainless steels predominate. Limits are normally placed on corrosion by-products detectable in the distilled water, primarily to ensure equipment longevity. Copper and iron are often measured as trace substances in the distillate, with limits of 0.02 mg/L. Parts of the seawater and brine systems of these large systems may be coated with epoxy or lined using various elastomers. Piping systems for seawater and brine are largely nonmetallic, with high-density polyethylene (HDPE) and glass-reinforced plastic, also known as *reinforced thermosetting resin pipe*, being the most common. Distilled water piping is generally stainless steel, polyvinyl chloride (PVC), or polypropylene, when the temperature is appropriate. Corrosion by-products are not typically monitored in the brine or cooling water systems because they are not expected to be present in large quantities in these sidestreams of the thermal desalination process.

2.5.2 MED Desalination

MED was originally developed in the 1830s, but more recently, in the 1960s, it was reconfigured to use thin-film evaporation techniques, making it more practical than the original concept. In the most common process configuration, the saline water to be evaporated is distributed over the outer surface of heated tubes. Within each MED effect, cool seawater is sprayed over a heat exchanger tube bundle, while steam flowing through the tubes is condensed into pure product water. Outside the tubes, the thin seawater film boils as it absorbs heat from the steam. The resulting vapor passes through mist eliminators to catch entrained brine droplets before the vapor is introduced into the tubes in the next effect. The process is repeated through the length of the plant. Alternative MED configurations that employ vertical tubes or plate-type heat transfer surfaces are also available.

Similar to the MSF process, a venting system must be incorporated within the MED system to ensure that the NCGs are continually swept away from the heat transfer surfaces. The venting systems usually use a combination of steam jet ejectors and small condensers to completely evacuate the NCGs to the atmosphere.

MED-TC combines a conventional MED system with a steam jet ejector or thermocompressor device. The thermocompressor makes use of the pressure of the motive steam to recycle the heat content, or enthalpy, of the process vapor from the final effect. By boosting the pressure of this process vapor, it can be recycled as heating steam in the first effect. Steam with a pressure greater than 100 kPa can be used in a thermocompressor. MED-TC systems differ from both MSF and MED systems, because unless special steps are taken to isolate it, the distillate can include steam condensate. This is an important consideration for drinking water systems, because steam condensate may include toxic boiler treatment chemicals. Additionally, even though MED-TC may use high-pressure and high-temperature steam, the maximum temperature within the process is rarely higher than 75°C. Both MED and MED-TC systems have an advantage over MSF, in that the clean vapor side of the process is at a higher pressure than the saline side; any loss of integrity of the heat transfer surface will lead to a generally small loss of product, not contamination of all of the product water.

Large systems produce between 5,000 and 25,000 m^3 of water per day from each plant with facilities having 4–10 or more machines. Small systems of 250–5,000 m^3/day are commonly used for water supply of industrial facilities and small communities. Typically, the feedwater undergoes only relatively coarse screening as a pretreatment. The cooling water returned to the source is generally 8°C–12°C warmer than the feedwater, and it is not deaerated, but it may contain small quantities of corrosion by-products or DBPs. It is possible with MED or MED-TC to configure the process to function with less cooling water, resulting in a higher temperature rise than with MSF. If environmental limits are not imposed and applied, this process can function with a temperature rise of over 20°C.

A portion of the cooling water return is used for makeup to the desalination process. This makeup water flow rate is generally 3–4 times the desired production rate of distilled water. A scale inhibitor may be added to minimize the deposition of mineral scale within the system. The scale-inhibiting chemicals are usually phosphate, polyphosphate, or polymaleic acids and are dosed at 1–5 mg/L. Periodically, antifoaming surfactants may be added to counteract seasonal changes in the feedwater. The most common antifoaming agent is a polypropylene oxide/polyethyleneoxide surfactant which is added at rates of 0.1–0.2 mg/L. In the process, the makeup water is heated to a TBT of between 60°C and 75°C before being boiled (in successively lower-pressure stages) to release water vapor, which can then be condensed and collected.

The separation process is primarily boiling evaporation with agglomeration systems (demisters) to collect and remove liquid carryover from the system. MED/MED-TC systems generally produce distilled water with a TDS content of between 20 and 50 mg/L. The distillate TDS comes from solids carryover from the separation process (or less likely from joint leakage or tube failure within the main or vacuum system condensers, allowing bypassing of the separation process). The integrity of the tubes and joints is not usually considered an issue with most MED/MED-TC configurations, because the distilled water side of the process is generally at a higher pressure than the saline side. The process releases dissolved gases from the seawater; these include air and carbon dioxide. The gases are vented to the atmosphere from the process (otherwise they would impede heat transfer) and are generally not considered harmful. These gases are also redissolved in the distillate, as predicted by Henry's law. The primary impact of this is the reduction of pH to around 6.2 as a result of high carbon dioxide.

The concentrate, frequently referred to as brine, is returned to the sea. Most commonly the brine is blended with cooling water from the MED/MED-TC process (as well as cooling water from other processes, if available). The brine will contain scale inhibitor and antifoam chemicals and is generally considered deaerated (prior to blending with cooling water). The brine TDS concentration is typically 1.4–1.8 times higher than that of the raw seawater or makeup water. The brine temperature is usually 15°C higher than that of the seawater before blending with cooling water.

2.5.2.1 MED Issues and Considerations

The cooling water returned to the sea is generally 8°C–12°C warmer than the ambient conditions. It is not normally deaerated but may contain small quantities of corrosion by-products or DBPs. MED systems generally produce distilled water with a TDS concentration between 20 and 50 mg/L. The distillate TDS content typically originates from spray carryover or from joint leakage or tube failure. NCGs are redissolved in the distillate, which results in the reduction of pH to around 6.2 (as a result of carbon dioxide). Volatile materials in the feedwater source are also distilled.

The feedwater concentrate (brine) is returned to the source (usually the sea). Most commonly, the brine is blended with cooling water from the MED process (as well as cooling water from other processes, if available). The brine typically contains scale inhibitor and antifoam chemicals and is deaerated (prior to blending with cooling water). The brine TDS concentration is generally 1.4–1.8 times higher than that of the raw seawater or makeup water. The brine is usually 5°C–25°C warmer than the seawater (before blending with cooling water).

There is no direct contact between the heating steam system and the MED desalination process; however, there is direct contact in the MED-TC process. Steam condensate does not mix with the seawater, brine, or distilled water. In systems that utilize steam that may contain low-grade treatment chemicals, a steam transformer or reboiler must be incorporated into the design. Condensate drains from the vacuum system could also contain dissolved or absorbed contaminants from feedwater decomposition and/or from steam. These drains sometimes connect into the brine side of the process (for some minimal heat recovery prior to rejection from the process). However, they can also be directly rejected if required.

As with MSF facilities, these systems can be manufactured from a variety of materials, but they are usually constructed from copper, nickel, and various molybdenum-bearing grades of austenitic stainless steels. Limits are normally placed on corrosion by-products detectable in the distilled water, primarily to ensure equipment longevity. Copper and iron are often chosen as trace substances in the distillate, with a limit of 0.02 mg/L each. Parts of the feed seawater and brine conveyance systems of these plants may be coated with epoxy or lined using various elastomers.

2.6 MEMBRANE DESALINATION

Membrane desalination is a process of separation of minerals from the source water using semipermeable or ion-selective membranes. Two general types of technologies are currently applied for membrane desalination: RO and electrodialysis (ED).

2.6.1 Desalination by Electrodialysis

In ED-based treatment systems, the separation between water and salt is achieved by passing a direct current through the ion-bearing water, which drives the ions in the source water through membranes to electrodes of opposite charge. A commonly used desalination technology that applies the ED principle is electrodialysis reversal (EDR). In EDR systems, the polarity of the electrodes is reversed periodically during the treatment process.

The ED-based systems use ion transfer (permselective) anion and cation membranes to separate the ions in the source water. These membranes are essentially ion exchange resin cast in sheet form of thickness 0.13–1.0 mm (typically 0.5 mm). The membranes allow a unidirectional transfer of ions of a given charge through them; for example, the cation transfer membranes allow only positively charged ions to pass. This process is shown in its simplest form in Figure 2.5. The water flows across, not through, the membrane surfaces, and ions are electrically transferred through the membranes from the source water stream to the concentrate stream under the influence of the direct current. Desalted water is produced in the compartments on the left of the anion transfer membrane (shown as AM on Figure 2.5) and concentrates in the compartments on the right.

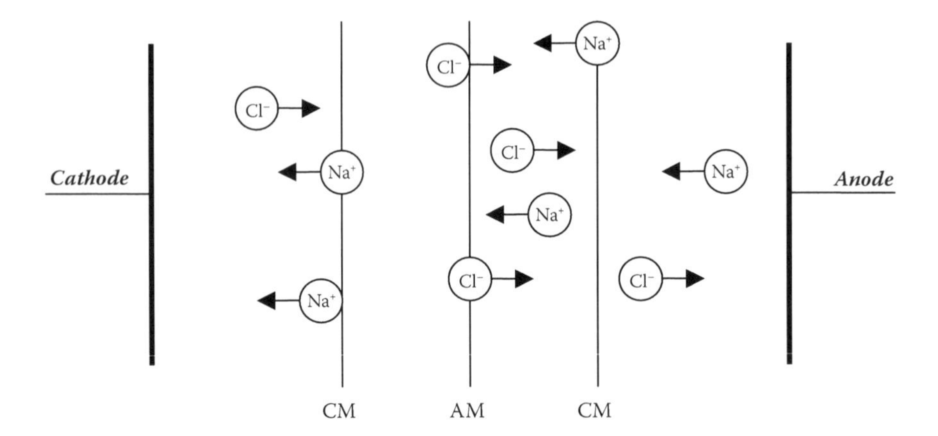

FIGURE 2.5 General schematic of an electrodialysis system (AM, anion transfer membrane; CM, cation transfer membrane).

The energy use for ED desalination is directly proportional to the amount of salt removed from the source water. TDS concentration and source water quality determine to a great extent which of the two membrane separation technologies (RO or ED) would be more suitable and cost-effective for a given application. Typically, ED membrane separation is found to be cost-competitive for source waters of TDS concentration lower than 3000 mg/L (FAO 2005). EDR is used instead of RO if higher recoveries are needed and the source water contains large amounts of scaling compounds, such as silica, that can precipitate and foul the RO membranes, if the source water may need to be oxidized (i.e., oxidation of arsenite to arsenate) or for source waters with very high biofouling potential.

2.6.2 Reverse Osmosis Desalination

RO is a process in which the product water (permeate) is separated from the salts in the source water by pressure-driven transport through a membrane. As a result of the RO process, desalinated water is transported under pressure through the membrane, while the minerals of the source water are concentrated and retained by the membrane. Application of high pressure for desalination is needed mainly to overcome the naturally occurring process of osmosis, which drives the desalinated water back through the membrane into the water of more concentrated mineral content. Nanofiltration (NF) is a process similar to RO, where membranes with order-of-magnitude-larger pore size are used to remove high-molecular-weight compounds and polyvalent ions that cause water hardness (e.g., calcium and magnesium).

Membrane desalination processes use semipermeable membranes—with pumping pressure as the driving force—to separate a saline feedwater into two streams: a low-salinity product (permeate) and a high-salinity stream (concentrate or reject). Although not technically correct, the concentrated reject stream is often referred to as brine. RO is not a true filtration process, because dissolved salts and other matter are not removed solely because of their size. This process relies on the ability of water molecules to diffuse through the membrane more readily than salts and higher-molecular-weight compounds. The membrane is the heart of an RO or NF system, and it is the barrier by which dissolved solids separation is accomplished. The membranes are made of long-chain high-molecular-weight organic polymers, which have an affinity for water. This hydrophilic characteristic allows water molecules to readily diffuse, or permeate, through the membrane structure, while restricting the passage of other substances. Three characteristics are commonly used to describe membrane performance: flux, salt rejection, and recovery:

- A membrane's *permeate flux* is the rate at which it will diffuse water molecules and is dependent on the membrane characteristics (i.e., thickness and porosity) and system operating conditions (i.e., feed pressure, temperature, and salt concentration). It is expressed in terms of flow rate per unit area (i.e., L/h per square meter). Similarly, the rate at which salts or other solutes diffuse through a membrane is referred to as a *salt flux*. The salt flux is proportional to the concentration difference between the feed and permeate solutions.
- *Salt rejection* refers to the effectiveness of a membrane at removing salts from a solution. The salt rejection varies slightly for specific ions, and the total rejection is determined by dividing the difference in feed and product water salt concentrations by the feedwater concentration.
- The proportion of feedwater that is recovered as product water is referred to as the *permeate recovery* or *conversion*. As a system's recovery rate increases, so does the reject stream salt concentration and, consequently, the osmotic pressure.

RO modules (units) can be arranged in various arrays to satisfy different application requirements. The simplest configuration is a single module. As the capacity of the system increases, additional modules can be added in parallel. The system performance in terms of product water quality and recovery ratio will be essentially identical to that of a system having a single module. To obtain higher recovery ratios or to produce higher-quality product water, it is often possible to arrange a system so that the concentrate or product from one stage becomes feedwater for a subsequent stage.

RO membrane desalination plants include the following key components: source water intake system, pretreatment facilities, high-pressure feed pumps, RO membrane trains, energy recovery, and a desalinated water conditioning system. The source water intake system could be an open surface water intake or series of seawater beach wells or brackish groundwater wells. Depending on the source water quality, the pretreatment system may include one or more of the following processes: screening, chemical conditioning, sedimentation, and filtration. Figure 2.6 shows a typical configuration of a seawater RO membrane system. The filtered water produced by the plant's pretreatment system is conveyed by transfer pumps from a filtrate water storage tank through cartridge filters and into the suction pipe of the high-pressure RO feed pumps. The cartridge filters are designed to retain particles

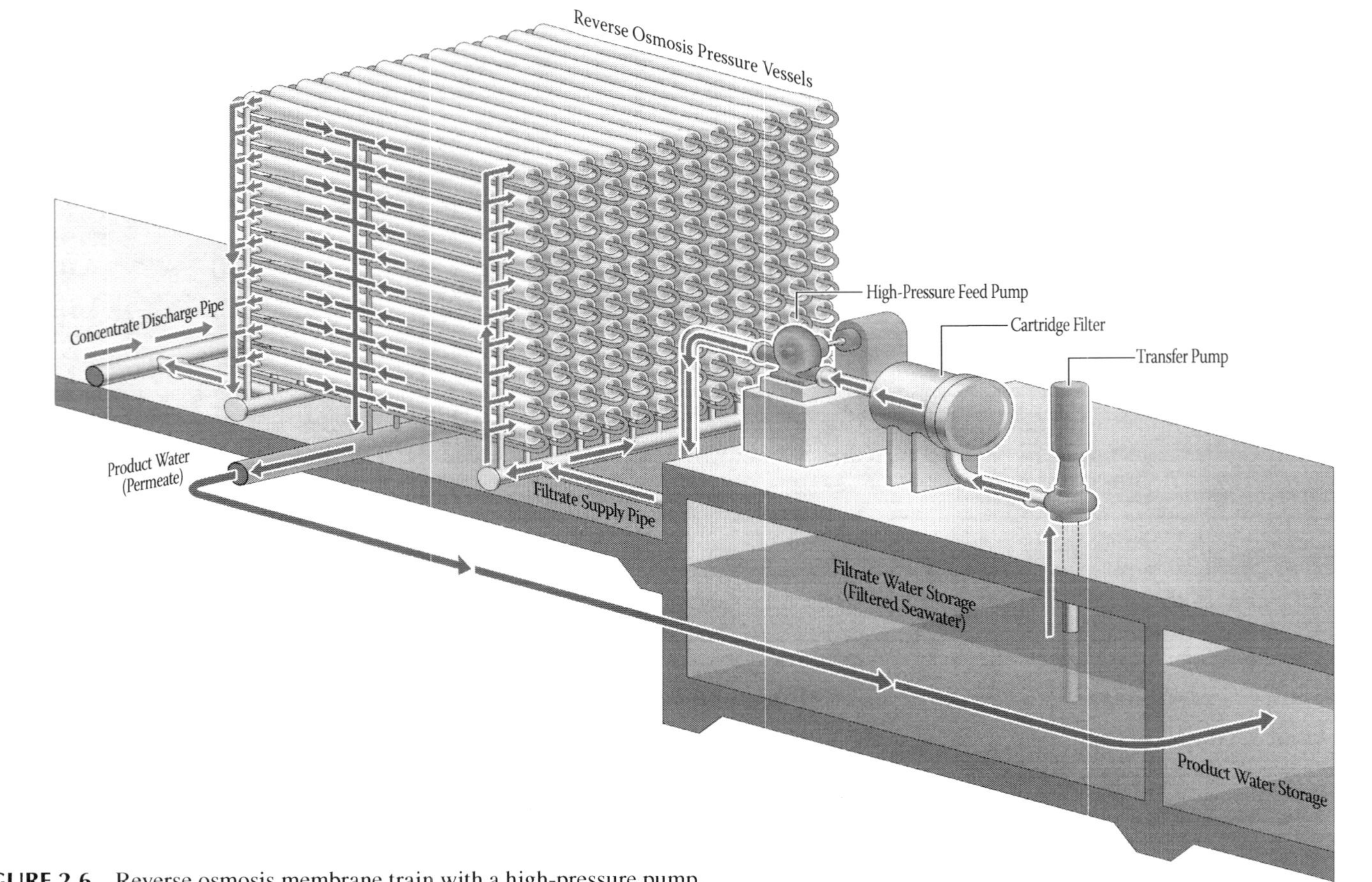

FIGURE 2.6 Reverse osmosis membrane train with a high-pressure pump.

of 1–20 μm that have remained in the source water after pretreatment. The main purpose of the cartridge filters is to protect the RO membranes from damage. The high-pressure feed pumps are designed to deliver the source water to the RO membranes at a pressure required for membrane separation of the fresh water from the salts (typically 15–35 bars for brackish source water and 55–70 bars for seawater). The actual required feed pressure is site specific and is determined mainly by the source water salinity and the configuration of the RO system.

The most widely used type of RO membrane elements consists of two membrane sheets glued together and spirally wound around a perforated central tube through which the desalinated water exits the membrane element. The first membrane sheet, which actually retains the source water minerals on one side of the membrane surface, is typically made of thin-film composite polyamide material and has microscopic pores that can retain compounds of size smaller than 200 Da. This sheet, however, is usually less than 0.2 μm thick; in order to withstand the high pressure required for salt separation, it is supported by a second thicker membrane sheet, which is typically made of higher-porosity polysulfone material that has several orders-of-magnitude-larger pore openings.

The commercially available RO membrane elements are of standardized diameters and length and salt rejection efficiency. For example, the RO membrane elements most commonly used for drinking water production in large-scale plants have 20 cm diameter and 100 cm length and can reject 99.5% or more of the TDS in the source water. Typically, SWRO membranes provide complete rejection of algal toxins, such as saxitoxin and domoic acid, as well as most human-made source water pollutants (pharmaceuticals, cosmetic products, etc.). Chapter 4 provides analysis of the microbiological aspects of membrane desalination and its efficiency in terms of pathogen removal.

These membranes can typically achieve over 4 logs of removal of *Giardia*, *Cryptosporidium*, viruses, and other pathogens in the source water. Desalination technologies are usually more effective than conventional water treatment technologies (sedimentation, filtration, chlorination) in removing pathogens and human-made pollutants from source water. In addition, typically, seawater and source water from deep brackish aquifers are usually less polluted than many surface water resources (rivers, lakes, or estuaries) in urbanized areas.

Standard membrane elements have limitations with respect to a number of performance parameters, such as feedwater temperature (45°C), pH (minimum of 2 and maximum of 10), SDI (less than 4), chlorine content (not tolerant to chlorine in measurable amounts), and feedwater pressure (maximum of 80–100 bars).

2.6.3 Membrane Issues and Considerations

2.6.3.1 Pretreatment

An inadequate pretreatment system may severely impact RO system performance. In fact, most RO system performance challenges are usually the result of a poorly designed/performing pretreatment system. It is of critical importance that the pretreatment system performance be satisfactory for the service intended.

2.6.3.2 Salt Rejection by RO Systems

Because salt rejection varies for different ions (generally, multivalent ions are rejected better than monovalent ions) and may be affected by factors such as temperature, it is important that the system design take into consideration all chemical constituents of the feedwater. Some constituents such as boron must be carefully considered under the full range of operating conditions to ensure that the maximum permeate levels are not exceeded. In addition, the membrane's salt rejection deteriorates over time, and it is standard practice to design a system based on the membrane's projected performance at the end of its normal operating life (3–5 years).

2.6.3.3 Salt Removal by ED-Based Systems

The TDS removal efficiency of ED desalination systems is not affected by nonionized compounds or those with a weak ion charge (i.e., solids particles, silica, organics, and microorganisms). Therefore, the ED membrane desalination processes can treat source waters of higher turbidity and biofouling and scaling potential than RO systems. However, the TDS removal efficiency of ED systems is typically lower than that of RO systems (15%–90% versus 99%–99.8%), and they do not remove pathogens and silica from the source water. Because of these key reasons, the ED-based systems have found practical use mainly for brackish water desalination and wastewater reuse for irrigation. The ED technology, in combination with ion exchange systems, is also frequently used for the production of ultrapure water for industrial applications.

2.6.3.4 Membrane Integrity

Inorganic fouling, biofouling, and scaling of membranes may be so severe as to result in their premature failure. In addition, manufacturing defects or mechanical problems with membranes or O-ring connectors can allow passage of dissolved solids, which could contaminate the permeate. It is important that an appropriate permeate monitoring and/or testing system be employed to detect excursions in quality indicating loss of integrity.

2.6.3.5 Hydrolysis

Some RO membranes are made of cellulose acetate and are susceptible to chemical decomposition or hydrolysis if the feedwater is operated outside a pH range of 3.5–7.5. Such systems should use acid injection to maintain an adequate pH to obtain a minimum 3-year membrane life.

2.6.3.6 Energy Recovery Cross-Contamination

RO energy recovery devices use the hydraulic pressure of the brine to pressurize feedwater. With some devices, it may be possible for the brine to leak through to the feedwater side of the device, increasing feedwater TDS. Adequate precautions should be taken to ensure that if such an event occurs, it does not increase feedwater TDS to a point that could prove detrimental to the RO operation.

2.7 POSTTREATMENT

Product water from desalination plants is characteristically low in mineral content, alkalinity, and pH. Therefore, desalinated water must be conditioned (posttreated) prior to final distribution and use. Typically, posttreatment of product water includes one or more of the following processes:

- Stabilization by addition of carbonate alkalinity
- Corrosion inhibition
- Remineralization by blending with high mineral content water
- Disinfection
- Water quality polishing for enhanced removal of specific compounds (e.g., boron, silica, dimethylnitrosamine [NDMA])

Posttreatment of permeate produced by the desalination system is needed for disinfection and mineral addition to protect public health and to safeguard the integrity of the water distribution system. In some cases, the same posttreatment process and conditioning chemicals allow achievement of both goals. For example, addition of calcium and magnesium salts to permeate stabilizes the product water and thereby protects the water distribution system against corrosion and also provides some essential minerals for reconstitution.

Usually, the ultimate application dosage of any multipurpose chemical is determined by the minimum dosage needed to achieve all purposes for which the conditioning chemical is added. If the use of the same chemical is not found to be cost-effective to achieve both the public health and the corrosion protection goals, then a combination of chemicals that yield the lowest overall cost of water production may be used to meet all posttreatment goals. For example, typically, calcium hypochlorite addition for disinfection meets the public health goals for pathogen inactivation and also adds some calcium. Therefore, calcium hypochlorite addition is typically combined with the feed of corrosion inhibitor to the desalination plant permeate, thereby achieving both public health and corrosion protection goals at a minimal life cycle cost.

2.7.1 Stabilization by Addition of Carbonate Alkalinity

The lack of carbonate alkalinity makes permeate from desalination plants very unstable and prone to wide variations in pH due to the low buffering capacity. Lack of carbonate alkalinity and calcium may also contribute to increased corrosion, as protective calcium carbonate films cannot be deposited on pipe walls. In existing systems, deficient carbonate alkalinity may cause previously deposited calcium carbonate films to dissolve. Monovalent ions such as chlorides as well as gases such as hydrogen sulfide, oxygen, ammonia, and carbon dioxide may pass through RO membranes to a greater degree than other ions or molecules; thus, they can also contribute to corrosion potential (Seacord et al. 2003).

TABLE 2.4
Factors Affecting Corrosion of Desalinated Water

Water Quality Parameter	Significance
pH	Low pH (typically below 7.0) may increase corrosion rates
	High pH (typically above 8.0, but not excessively) may reduce corrosion rates
Alkalinity	Provides water stability and prevents variations in pH
	May contribute to the deposition of protective films
	Highly alkaline water may cause corrosion in lead and copper pipes
Calcium	May deposit as a calcium carbonate film on pipe walls to provide a physical barrier between metallic pipe and water
	Excess concentrations may decrease the water transmission capacity of pipes
Hardness	Hard water is generally less corrosive than soft water if calcium and carbonate alkalinity concentrations are high and pH conditions are conducive to calcium carbonate deposition
Chlorides	High concentrations may increase corrosion rates in iron, lead, and galvanized steel pipes
Silica	Can react to form a protective film when present in dissolved form
Phosphate	Can react to form protective film
Temperature	Can impact solubility of protective films and rates of corrosion

Source: Courtesy of N. Voutchkov.

Corrosion can affect many aspects of drinking water supply, including pumping costs, public acceptance of treated water, disinfection efficacy, and public health due to exposure to heavy metals (e.g., lead, copper, and cadmium). In conventional water treatment, corrosion is defined as the degradation of pipe materials due to a reaction with water. This reaction can be physical, chemical, and electrically or biologically induced. Similarly, in a desalinated water supply, chemical reactions can cause the degradation of metallic pipe that comes in direct contact with water. Chemical reactions that cause corrosion of metallic surfaces are affected by many water quality parameters, including pH, alkalinity, calcium, hardness, chlorides, silica, phosphate, and temperature. The significance of each of these parameters is summarized in Table 2.4. Although many other water quality parameters, such as TDS, dissolved oxygen, hydrogen sulfide, sulfate, and biological activity, affect corrosion as well, these parameters are not the emphasis of this chapter, as they are not unique constituents of desalinated water.

2.7.2 Corrosion Indices

Water stability and corrosion potential may be characterized by parameters (corrosion indices) indicating the potential of the desalinated water to precipitate calcium carbonate and by parameters that address corrosivity caused by specific compounds in permeate. Calcium-carbonate-based corrosion (stability) indices include the Langelier saturation index (LSI), the calcium carbonate precipitation potential (CCPP), and the aggressiveness index (AI). The Larson ratio (LR) is a corrosion index that does not predict calcium carbonate deposition, but rather presents a qualitative relationship between molar concentrations of chloride and sulfate to bicarbonate alkalinity concentrations and relates that ratio to corrosion potential.

2.7.2.1 Langelier Saturation Index

The LSI is a qualitative assessment of water's potential to precipitate calcium carbonate. This index is based on the difference between the pH of the unconditioned permeate and the pH of the permeate when it is just saturated with calcium carbonate. A negative LSI indicates that the water is undersaturated (i.e., calcium carbonate will dissolve), whereas a positive LSI may indicate that the water is oversaturated (i.e., calcium carbonate will precipitate). However, this index does not actually account for the amount of carbonate in water. Therefore, while calculations may indicate a positive LSI, it is possible that very little calcium carbonate may precipitate.

Variations of the LSI calculation exist. The original LSI was developed to predict the potential precipitation of calcium carbonate in a specific fresh surface water source rather than for desalination permeate, and it has limited suitability for assessment of the corrosion potential of desalinated water (permeate). The LSI is also subject to misinterpretation, as it overestimates saturated conditions at high pH when additional carbonate may not actually be available for precipitation, which may be interpreted as highly scaling conditions when water is actually undersaturated (Seacord et al. 2003).

2.7.2.2 Calcium Carbonate Precipitation Potential

As opposed to the other carbonate-based corrosion indices, the CCPP is an index that quantifies the amount of calcium carbonate that may dissolve or precipitate. The CCPP is a true indication of water's potential to deposit calcium carbonate, thus forming a protective coating on pipe walls to inhibit corrosion. This index is the most useful tool for developing corrosion control and posttreatment strategies for desalinated seawater. Positive values of the CCPP indicate the concentration of calcium carbonate that exceeds the saturated condition, whereas negative values denote the amount of calcium carbonate that must dissolve to reach a saturated condition. The CCPP is a function of alkalinity and is defined and discussed in detail in Rossum and Merrill (1983) and Seacord et al. (2003). A recommended range for the CCPP in desalinated water is between 4 and 10 mg/L as calcium carbonate. Higher values can be acceptable when combined with the use of calcium complexing agents, such as polyphosphate-based scale inhibitors. However, when the CCPP is too high and scale inhibitors are not used, the water transmission capacity of a pipeline may be reduced due to excessive calcium carbonate precipitation. The CCPP has not been widely used by water treatment engineers owing to the analytical complexity of hand calculation. Spreadsheet models can be used to calculate the CCPP for a wide range of water qualities (AWWA 1996).

2.7.2.3 Aggressiveness Index

The AI was developed for asbestos–cement pipe with water temperatures ranging from 4°C to 27°C. The AI is calculated as a function of pH, calcium concentration, and alkalinity (Schock 1991). Waters with AI less than 10 are considered highly aggressive; waters with AI between 10 and 12 are considered mildly aggressive; and waters with AI greater than 12 are considered noncorrosive and depositing.

The AI has significant limitations for predicting corrosion potential in asbestos–cement pipe. AI is only an approximation of the LSI and does not indicate the true potential for calcium carbonate precipitation. Similar to LSI, AI also overestimates saturation at high pH when conditions are undersaturated. Additionally, at a pH of 8–9, when CCPP and LSI indicate saturated conditions, the AI underestimates saturated conditions, which further indicates the limitation of this index and its application to the development of corrosion control measures for desalinated seawater.

2.7.2.4 Larson Ratio

Research completed by Larson (1970) demonstrated that chloride and sulfate can increase both corrosion rates and iron concentrations in water when water is conveyed in ferrous metal and alloy pipes. It is important to understand the effect of chloride, especially when designing desalination systems that treat high-chloride brackish waters and seawater. Pipes used to convey water to the desalination system and permeate piping must be compatible with high chloride concentrations. PVC is typically used for low-pressure piping, whereas high molybdenum content (6%–8% or higher) stainless steel alloys are often used for high-pressure piping. However, corrosion in distribution systems cannot be prevented by material selection alone. Posttreatment is still required, as distribution piping may include ferrous metals that are prone to chloride attack.

It is recommended that the LR be maintained at less than 5 for municipal applications. The LR is a ratio of the molar concentrations of chloride and sulfate to bicarbonate alkalinity. This index accounts for the fact that although desalinated water is often higher in chlorides than in other water sources, typically it has an order-of-magnitude-lower content of sulfates, as well as the fact that chlorides and sulfates both have an effect on corrosion. It also indicates that the LR of the water can be reduced (i.e., corrosion rate can be diminished) by increasing the bicarbonate level in the water.

2.7.3 Corrosion Control Methods

Widely used corrosion control strategies include the placement of a physical barrier between water and metallic pipes to protect the pipe from the corrosive properties of water. Physical/chemical barriers include chemical precipitants such as calcium carbonate, silica, and orthophosphate. Predominantly, the precipitation of calcium carbonate on pipe walls has been used to control corrosion of desalinated drinking water. However, the amount of calcium carbonate deposited should be closely controlled to prevent pipes from losing their capacity to convey water because too much calcium carbonate is deposited. To develop posttreatment concepts for corrosion control following desalination, it is first necessary to have treatment goals. The following goals are used as a guide for developing posttreatment strategies:

- Alkalinity = ≥40 mg/L as calcium carbonate
- CCPP = 4–10 mg/L as calcium carbonate (without the addition of calcium complexing agents)
- LR = <5 (for steel pipelines only)
- LSI = +0.5 to +1.0
- Total hardness = >50 mg/L as calcium carbonate

Alkalinity greater than or equal to 40 mg/L as calcium carbonate is a prime goal, because experience shows that alkalinity less than this value may result in poor buffering and pH variations in water systems. CCPP is preferred as the carbonate-based scale formation index, as the other indices provide only a qualitative assessment of carbonate deposition potential.

2.7.3.1 Decarbonation

Decarbonation, or removal of excess carbonic acid or carbon dioxide, may be required, owing to the presence of high concentrations of carbonic acid, which is typically accompanied by lowering product water pH. The presence of excess hydrogen ion (i.e., low pH) has been linked to increased corrosion potential owing to the presence of an electron acceptor in corrosion reactions. Carbonic acid may result from the conversion of bicarbonate when acid is added to the desalination facility feedwater as a method of controlling calcium carbonate scaling on the RO membrane. Decarbonation will help increase the finished water pH.

Decarbonation consists of an air transfer process, where carbonic acid (carbon dioxide) is in the air phase. Packed tower aeration, tray aeration and, more recently, hollow fiber membrane aeration can be used for removing carbonic acid from desalination permeate. Decarbonation is typically used in combination with other posttreatment processes, as it may be beneficial to convert some carbonic acid back to bicarbonate alkalinity. Combined use of decarbonation with pH adjustment may be more economical, as this will help control the cost of chemicals used to increase pH while still producing the desired pH, alkalinity, and CCPP.

2.7.3.2 Addition/Recovery of Alkalinity

Carbonic acid is needed to maintain bicarbonate alkalinity, which in turn is critical for permeate stability and corrosion protection. Bicarbonates are buffers against significant pH variations; such buffering is critical to achieve adequate corrosion protection and disinfection. It may be necessary to supplement alkalinity with chemical treatment when sufficiently high concentrations of carbonic acid are not available in the desalination plant permeate to generate the desired amount of bicarbonate alkalinity with pH adjustment. Whereas hydroxide addition will increase the finished water alkalinity and pH, carbonate and bicarbonate alkalinity are required to produce a CCPP within the desired range; they also help provide buffering capacity, which in turn will prevent drinking water pH variation in the distribution system. The following posttreatment methods are widely used to add or recover alkalinity in desalination permeate:

- Addition of caustic soda or lime to permeate containing carbonic acid
- Addition of carbonic acid followed by the addition of caustic soda or lime
- Addition of sodium carbonate or sodium bicarbonate

Caustic soda or lime addition to permeate containing carbonic acid is referred to as *alkalinity recovery*. It may be necessary to combine alkalinity recovery with a decarbonation process to control chemical costs when carbonic acid concentrations are high. When desalination plant permeate water has a relatively low concentration of carbonic acid, carbonic acid can be added using a carbon dioxide gas feed system and then converting the carbonic acid to bicarbonate alkalinity. This can be done with carbon dioxide that has been recovered during the desalination process. Caution should be exercised to prevent the introduction of volatile organic compounds that may be present in the vent gases. An alternative method of increasing carbonate-based alkalinity is the addition of sodium carbonate or sodium bicarbonate. In practice, it may be necessary to monitor dose rates to make certain that finished water sodium concentrations are kept below the 200 mg/L limit that is suggested by WHO to avoid unacceptable taste (WHO 2008).

2.7.3.3 Addition of Hardness

A calcium carbonate film deposited on pipe walls can be used as a physical barrier to prevent corrosion. There are a variety of posttreatment methods used to add hardness back to desalination plant permeate. These may include the addition of lime or contact filtration through limestone (calcite or dolomite) filters. Slaked lime (calcium hydroxide) is added to permeate water to provide calcium and alkalinity (i.e., hydroxide alkalinity) as well as to adjust product water pH. When adding lime to desalination plant permeate, it is important to consider that the solubility of calcium carbonate is dependent on pH, temperature, and ionic strength. Lime may not dissolve easily, and a residual turbidity may result, which is a disadvantage of this approach. Posttreatment may require the addition of acid (e.g., carbonic acid) to help dissolve the lime and produce the desired hardness concentration and CCPP. If the permeate water is too warm, the rate of lime dissolution will also be slowed.

There are a few approaches available to encourage the dissolution of slaked lime in water with temperatures higher than 25°C. One approach is to provide multiple points for carbonic acid injection and a separate lime contact chamber that creates highly turbulent conditions and provides contact times of 5–10 min. Another approach used to enhance the reaction of relatively warm plant permeate with lime is mixing the lime suspension and plant permeate in the product water storage tank using large recirculation pumps. This approach is cost-effective only if the unit power cost is relatively low (i.e., $0.02–0.03/kWh).

Limestone filters have been used extensively in Europe and the Middle East, often in conjunction with carbonic acid addition to adjust pH, alkalinity, and CCPP, in order to add hardness and to produce finished water that is stable. Limestone (calcite or dolomite) pebbles are widely used for this application. While calcite pebbles provide only calcium hardness to the water, use of dolomite contributes both calcium and magnesium, which could be an advantage if the water is used for irrigation of certain crops or for nutrient embellishment of drinking water. Limestone filters combine two advantages: enhanced contact time and final filtration of the plant product water, allowing the controllable production of low-turbidity permeate. Limestone filter cells are usually concrete structures designed at loading rates several times higher than those used in conventional media filters. The limestone bed is typically 2.5–3 m deep. The process can be designed with or without carbon dioxide. The limestone in the bed dissolves as the desalination plant permeate passes through the media. Typically, when the limestone bed loss is between 10% and 15% of the original height, additional limestone is added. Usually, the cost per ton of limestone is higher than that of lime, and the use of limestone filters requires the construction of concrete filter cells and service facilities, which adds to the overall plant construction. Ultimately, the decision regarding use of lime feed system or limestone filters has to be based on a combination of life cycle costs, which greatly depends on the cost of lime, limestone, and facility construction costs, as well as operator skill level.

2.7.3.4 Corrosion Inhibition

Corrosion inhibitors are widely used to reduce the corrosivity of desalination plant permeate. Phosphate and silicate inhibitors form protective films on pipe walls that limit corrosion or reduce metal solubility. Orthophosphates react with pipe metal ions directly to produce a passivating layer. Silicate inhibitors form a glass-like film on pipe walls. Often, these inhibitors are added after corrosion has already occurred. In such cases, typical practice is to add the inhibitor at approximately three times the normal concentration for several weeks to begin the protective film formation. Initial doses should be continuous, and water circulation is required to completely distribute the inhibitor to all parts of the distribution system. Use of corrosion inhibitors instead of alkalinity addition is often more suitable when the water distribution system is made of nonmetallic piping (e.g., PVC, fiberglass, or HDPE pipe). In this case, the use of corrosion inhibitors avoids the potential problems that stem from the increase in product water turbidity associated with the addition of lime or other calcium-based minerals and reduces the overall chemical conditioning costs. The supporting documents referenced at the end of this chapter provide additional information on this topic.

2.7.3.5 Permeate Remineralization by Blending with Source Water

Permeate has a very low content of calcium, magnesium, and other minerals. Therefore, small amounts of minerals are often added to permeate by blending it with source water. This practice is frequently used for both brackish water plants and thermal desalination plants and is acceptable only when the source water is of high quality or pretreated appropriately for both microbial and chemical concerns and the blend meets all applicable water quality standards. When seawater is the source of the blending water, blending is limited to about 1% as a result of taste considerations. At a 1/99 blending ratio with seawater, this could add 4–5 mg calcium/L and 12–17 mg magnesium/L, plus sodium chloride, and other ions. Bromide in the seawater may lead to DBP and bromate issues in the finished water, so this approach is less common.

The source water used for blending must be treated prior to its mixing with the desalination plant permeate. The type and complexity of source water treatment depend on its quality. At a minimum, the source water used for permeate remineralization has to be filtered through cartridge filters and probably disinfected. Enhanced source water treatment such as granular activated carbon filtration is recommended for source water exposed to potential contamination from excessive algal growth, surface runoff, or other human-made sources of elevated organics or turbidity in the water. Pretreatment chemicals (such as acid) may need to be added, depending on where the blending source water is split from the feedwater piping. The LR should be checked in blending applications, especially if steel pipes are used for permeate conveyance.

2.7.4 Product Water Disinfection

Chlorine in various forms (e.g., sodium hypochlorite, chlorine gas) is generally used for disinfection because of its recognized efficiency as a disinfectant and because a reduced level of DBP precursors is generated in permeate. However, other final disinfectants, such as chlorine dioxide or even chloramines as a secondary disinfectant and ozone or ultraviolet (UV) light, could be used in combination with chlorination to control microbial regrowth, depending on specific conditions.

2.7.4.1 Chlorination

Chlorination is the most widely used disinfection method, and chlorine gas and sodium hypochlorite are the two most popular chemical forms. The desalination process has likely eliminated pathogens and other undesirable microorganisms. The typical target chlorine dosage that provides adequate disinfection depends on two key factors: permeate temperature and contact time. Usually, the chlorine dosage used for disinfection is 1.5–2.5 mg/L. Although very popular worldwide, the use of chlorine gas is associated with potential safety considerations associated with accidental gas releases. Therefore, chlorine gas disinfection facilities have to be equipped with gas detection, containment, and treatment facilities that provide adequate protection of public health. A 5%–15% solution of sodium hypochlorite is safer to use, handle, and store than chlorine gas. Sodium hypochlorite can be delivered to the desalination plant site as a commercial product, or it can be produced on site using low-bromide sodium chloride (salt). However, electrolysis of seawater to produce hypochlorite is not appropriate, because its high bromide content produces large amounts of bromate and probably brominated DBPs.

2.7.4.2 Chloramination

Chloramination is widely used principally as a secondary disinfectant because of its lower biocidal potency. This disinfection method includes sequential addition of chlorine and ammonia to the product water to form chloramines. Chloramines have a significantly slower rate of decay than free chlorine and therefore are often favored, especially for product water delivered to large distribution systems with high temperatures and long retention times and high potential for chlorine residual loss. Chloramination typically results in the creation of lower levels and different types of DBPs than with free chlorine disinfection. It may contribute to nitrite or NDMA production under some conditions. As desalinated seawater has very low amounts of organics, the use of chloramines for seawater desalination is not as advantageous as it may be for disinfection of drinking water produced from brackish or freshwater sources and therefore is not widely practiced. However, chloramination of desalinated water may be necessary if this water is planned to be blended with other water sources disinfected with chloramines. If chlorinated desalinated water is blended with chloraminated drinking water produced from a fresh surface water source, mixing of the two types of water may result in accelerated decay of chlorine residual of the blend of two waters. To avoid such decay, it is recommended that the desalinated water be chloraminated at higher dosages than the fresh source potable water with which it will be blended.

2.7.4.3 Chlorine Dioxide

Chlorine dioxide is widely used in preoxidation and postdisinfection of drinking water as an alternative to chlorine gas or sodium hypochlorite. Chlorine dioxide does not form significant quantities of total trihalomethanes (TTHMs) and adsorbable organic halogens (Molley and Edzwald 1988) in comparison with other chlorinated disinfectants, and bromate is not formed even if the desalinated water is blended with other sources containing bromide ion. The main chlorine dioxide by-products are chlorite and chlorate, together with some organic oxidation products. To minimize chlorate formation, it is necessary to improve the on-site generation of chlorine dioxide by using properly designed generators capable of producing very pure chlorine dioxide solutions and reaching very high conversion of the reagents (Gordon 2001). When chlorine dioxide undergoes chemical reduction in water treatment processes, about 60% is converted to chlorite ion. However, in desalinated water, the dosage of chlorine dioxide necessary to maintain a residual in the distribution network is quite low (<0.4 mg/L, which is about one-fourth of the required chlorine dosage) (Belluati et al. 2007), and therefore the chlorite residue is expected to be much lower than the current WHO limit (0.7 mg/L) (WHO 2008).

2.7.4.4 Ozonation

Ozonation is a widely accepted practice for the disinfection of product water from freshwater sources. However, ozonation of RO-desalinated water is associated with the potential formation of excessive amounts of assimilable organic carbon and bromate as a result of the relatively high content of bromide in the permeate compared with that in drinking water from other surface water sources. Additional information on this topic is provided by Perrins et al. (2006).

2.7.4.5 Ultraviolet Light Disinfection

UV irradiation of desalination plant permeate is a viable disinfection alternative that is particularly useful for *Cryptosporidium* oocyst inactivation. The disinfection of desalinated water will typically require lower UV dosages than those used for UV disinfection of other surface water sources because of the lower turbidity and lower content of pathogens in the desalinated water. Another advantage of UV disinfection is that it does not add any chemicals to the product water, and therefore the product water has a low content of DBPs. However, it also does not leave a disinfectant residual to control regrowth; disinfectant (chlorine or chloramine) can be added later.

2.7.5 Water Quality Polishing

Water quality polishing is used for enhanced treatment of specific compounds (e.g., boron, silica, NDMA) when these compounds have to be removed from the water to meet water quality targets for drinking or industrial use. Depending on the compounds targeted for removal, the treatment technologies may include ion exchange, granular activated carbon filtration, additional multistage or multipass membrane RO treatment, or a combination of treatment processes, which could include advanced oxidation. Parekh (1988) and Wilf et al. (2007) provide additional information on permeate polishing.

2.7.6 Posttreatment Issues and Considerations

2.7.6.1 Effect of Disinfection on Corrosion Control

It is important to consider the impact of disinfection processes on finished water pH and the resultant impact on the CCPP. Chlorine gas addition decreases pH and alkalinity as a result of the formation of hypochlorous acid, whereas sodium hypochlorite and calcium hypochlorite increase the pH and alkalinity of the product water.

2.7.6.2 Use of Sodium Hypochlorite Produced from Seawater

Sodium hypochlorite used for disinfection either can be delivered to the desalination plant as a commercial product or can be generated on-site using commercially available sodium chloride. The main advantage of on-site generation at the treatment plant, especially for large plants, is that it minimizes space requirements for storing large quantities of sodium hypochlorite solution, thus reducing hypochlorite solution strength decay and formation of chlorate during long storage. Usually, sodium hypochlorite solutions decay rapidly over time and lose 10%–20% of their strength over a period of 10–15 days, while chlorate concentrations increase, especially in warm climates. The rate of solution strength decay depends mainly on the initial concentration of the sodium hypochlorite, the ambient temperature, and the exposure to sunlight.

On-site sodium hypochlorite generation using commercially available high-grade low-bromide sodium chloride instead of seawater is recommended. Although the use of seawater as a source of chloride for the sodium hypochlorite generation process is less costly and simpler, it results in the generation of higher concentrations of DBPs and bromate because of the naturally high level of bromide in the seawater. When blended with desalination plant permeate, the sodium hypochlorite generated from seawater increases the concentration of DBPs and bromate in the drinking water.

2.7.6.3 Chloramination and Total Chlorine Residual Stability

Disinfection of desalinated seawater with chlorine results in a stable and long-lasting chlorine residual that provides adequate residual disinfection in the distribution system in many conditions. However, ammonia addition to permeate to form chloramines may result in an accelerated decay of the total chlorine residual if the bromide concentration of the permeate is above 0.4 mg/L (McGuire Environmental & Poseidon Resources 2004). Applying a combination of chlorine and ammonia to desalinated water with bromide levels above 0.4 mg/L may yield an unstable residual that decays at an accelerated rate because of the rapid conversion of chloramines to bromamines, which are much more chemically reactive. The destabilizing effect of bromide on the chloramination process can be mitigated by superchlorination (i.e., applying initial chlorine to permeate at dosages of 3.0–4.0 mg/L) or by producing permeate with a bromide level below 0.4 mg/L. The former would increase by-products, including organobromines.

2.7.6.4 Ozonation and DBP and Bromate Formation

Ozonated permeate may contain high concentrations of bromate. This could be addressed by reducing the bromide level before ozonation, which is costly. Ozonation does not leave a disinfectant residual to suppress the regrowth of microorganisms and biofilms.

2.7.6.5 Blending and Compatibility of Desalinated Water with Other Water Sources

Usually, blending desalinated water with surface water or groundwater of elevated salinity has a very positive effect on the quality of the water blend and is therefore highly desirable. Blending of low-DBP desalinated seawater with surface water with high-DBP content can reduce the overall DBP levels of the drinking water. However, when desalinated water has a high content of some minerals, such as bromide, boron, sodium, chlorides, calcium, and magnesium, the blending of this desalinated water with drinking water produced from other sources (river, lake, or groundwater) may have a negative effect on the blended water quality. Therefore, the compatibility of the various water sources must be taken into consideration prior to their blending. Specific issues that must be investigated include the following:

- Bromide and TOC concentrations in the various waters and their effect on the DBP formation and concentration in the blend.
- Type of disinfection used for the various water sources and the effect on DBP formation and chlorine residual stability.
- Sodium and chloride levels in various sources—desalination plant permeate may have higher sodium levels compared with other freshwater sources.
- Temperature of various water sources—the higher temperature may result in accelerated nitrification and corrosion in the distribution system; in contrast, the desalinated water is low in organics, and the net effect may be positive and may negate the temperature effect on the growth rate of the nitrifying organisms.

- Calcium and magnesium in the various sources—before blending, desalinated water usually has significantly lower levels of calcium and magnesium compared with fresh surface water sources. Blending of desalinated water with drinking water of high hardness may be sufficient to provide the needed water stability if the resulting blend meets the target water quality requirements for corrosion control described in Section 2.7.3. Conveyance of desalinated water in long pipelines may pose chlorination and corrosion problems, especially in warm climates. Loss of disinfectant residual is one of the main challenges in such systems, mainly because of residual disinfectant decay caused by high-temperature water. This may be addressed by superchlorination (i.e., feeding chlorine at dosages resulting in breakpoint chlorination, i.e., 3.5–4 mg/L); reinjection of chlorine along the length of the pipeline, which could be activated when the chlorine residual drops below 0.5 mg/L; or use of chloramine instead of chlorine for disinfection because of its lower chemical reactivity. Chloramines have slower decay rates than free chlorine.
- Loss of calcium alkalinity, which may occur over the length of the pipeline and thereby may result in corrosion problems. There are several alternatives to address this challenge: (1) reinject calcium-conditioning chemicals or corrosion inhibitors along the pipeline route at locations where the water LSI is reduced to a negative level; or (2) use nonmetal pipeline materials such as HDPE that are not sensitive to low levels of calcium alkalinity in the water.
- Disturbance of the steady-state condition in water in an existing system by introducing desalted water of a notably different character.

2.8 CONCENTRATE MANAGEMENT

One of the key limiting factors for the construction of new desalination plants is the availability of suitable conditions and locations for disposal of the high-salinity sidestream commonly referred to as *concentrate* or *brine*.

2.8.1 Concentrate Characterization and Quality

Concentrate is generated as a by-product of the separation of the minerals from the source water used for desalination. This liquid stream contains most of the minerals and contaminants of the source water and pretreatment additives in concentrated form. The concentration of minerals and contaminants in the concentrate is usually 2–10 times higher than that in the source water, depending on the recovery of the desalination plant. If chemical pretreatment is used, such as coagulants, antiscalants, polymers, or disinfectants, some or all of these chemicals may reach or may be disposed of along with the plant discharge concentrate.

The quantity of the concentrate is largely a function of the plant recovery, which in turn is highly dependent on the TDS concentration of the source water. Usually, brackish water desalination plants and water reclamation plants operate in a recovery range of 65%–90%, depending on the influent TDS concentration. Seawater desalination plant recovery is typically limited to 40%–65%. NF plants primarily reject divalent ions (e.g., calcium, magnesium, and sulfate), and therefore the concentrate from these plants is of lower sodium and chloride content than that of BWRO and SWRO plants. The TDS level of concentrate from seawater desalination plants is usually in a range of 65,000–85,000 mg/L, whereas that from brackish water plants may vary between 1,500 and 25,000 mg/L. The concentration of particles, TSS, and biochemical oxidation demand (BOD) in the concentrate is usually below 5 mg/L, because these constituents are removed by the plant's pretreatment system. However, if plant pretreatment waste streams are discharged along with the concentrate, the blend may contain elevated turbidity, TSS, and occasionally BOD. Acids and scale inhibitors added to the desalination plant source water will be rejected in the concentrate and will affect its overall mineral content and quality. Often, scale inhibitors contain phosphates or organic polymers. The concentration of scale inhibitors may reach 20–30 mg/L.

An important unique aspect of the thermal desalination plant discharge is that it has an elevated temperature and therefore is a source of thermal pollution, whereas the temperature of the discharge of a typical membrane seawater desalination plant is similar to that of the ambient ocean water. This is important, as the energy dissipated in the environment (i.e., the thermal pollution load) of a thermal desalination plant is very high; as can be seen from Table 2.5, it is substantially higher than the thermal discharge associated with a power plant of similar size. The thermal discharge load of a thermal desalination plant could be reduced by increasing the MSF and MED plant performance ratio.

Figure 2.7 shows the amount of thermal energy dissipated to the sea through the heat reject section of an MSF plant against the plant performance ratio (PR) at various distiller capacities. The figure clearly indicates that the thermal discharge from an MSF plant can be decreased significantly if the plant PR ratio is increased.

As seen in Figure 2.7, an MSF plant with a capacity of about 45,000 m^3/day and a PR of 9 dissipates the same energy in the sea as an MSF plant of about 30,000 m^3/day and a PR of 7. New thermal desalination plants are generally specified with high PRs

TABLE 2.5
Environmental Impacts of Power Generation and Desalination Processes

Reference Process	Type of Process	Energy Dissipated in the Environment (Mw)	TDS Increase (%) With Respect to the Uptake
Power generation	Conventional cycle	50	0
(150 MW)	Combined cycle	10	0
Desalination plant	MSF (PR 9)	120	15–20
(33,000 m^3/day)	MED (PR 9)	100	15–20
	SWRO	0	50–80

Source: Courtesy of N. Voutchkov.

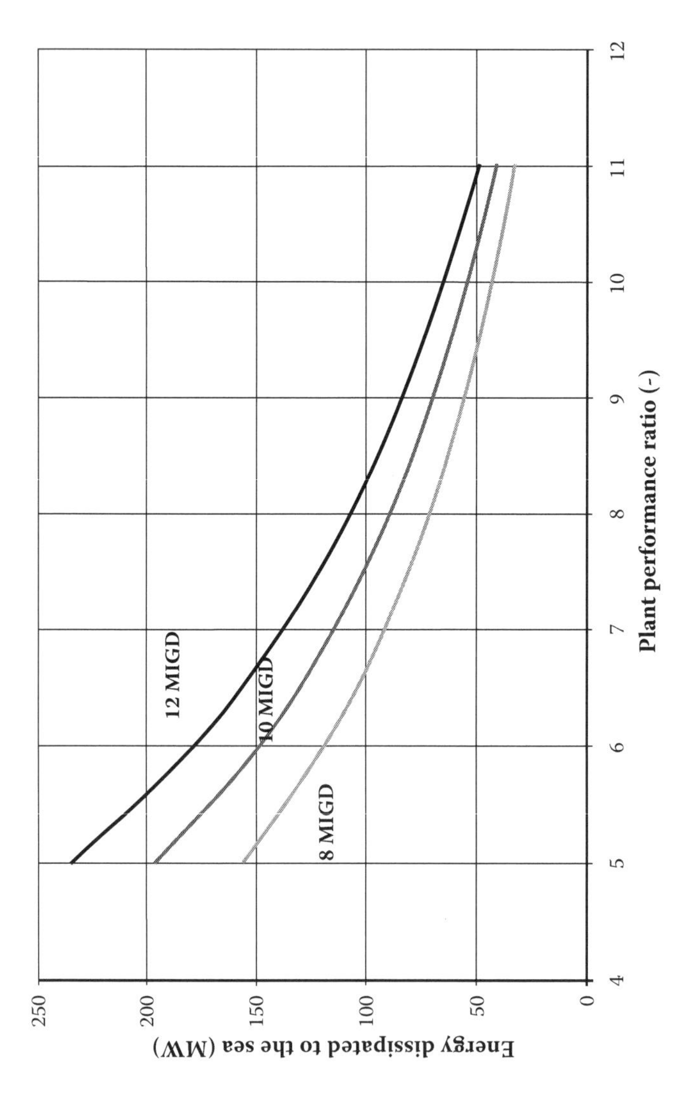

FIGURE 2.7 Thermal energy discharge load of MSF plants.

because this also decreases carbon dioxide and nitrogen oxide emissions. However, there are a number of old installations operating with a PR of 5. The thermal load discharge issue is an important issue, especially for large thermal desalination plants.

2.8.2 Overview of Concentrate Management Alternatives

Chapter 6 of this book provides guidance for the preparation of environmental impact assessments of desalination plants, part of which deals with concentrate management. The following sections discuss the most widely used concentrate disposal alternatives. These include

- Discharge to surface waters
- Discharge to sanitary sewer
- Deep-well injection
- Evaporation ponds
- Spray irrigation
- Zero-liquid discharge

According to a study by the U.S. Bureau of Reclamation (Mickley 2006), the concentrate disposal methods most widely used in the United States are those shown in Table 2.6. These results are based on a survey of 203 desalination plants completed in the year 2000. The survey included only plants with a capacity larger than 200 m^3/day. Approximately 95% of the surveyed plants were NF, BWRO, or SWRO facilities. Table 2.6 reflects typical concentrate management practices worldwide.

In addition to the listed methods, two additional trends in concentrate management are gaining acceptance worldwide:

- Regional concentrate management
- Beneficial concentrate use

A brief description of each of the concentrate disposal methods and key issues, considerations, and mitigation measures associated with their implementation are summarized in the following text.

TABLE 2.6
Concentrate Disposal Methods and Their Frequency of Use

Concentrate Disposal Method	Frequency of Use (% of Surveyed Plants)
Surface water discharge	45
Sanitary sewer discharge	27
Deep-well injection	16
Evaporation ponds	4
Spray irrigation	8
Zero-liquid discharge	0

Source: Courtesy of N. Voutchkov.

2.8.3 Concentrate Discharge

2.8.3.1 Discharge of Concentrate to Surface Waters

This disposal method involves the discharge of the desalination concentrate to a surface water body such as the nearby ocean or sea, river, estuary, bay, or lake. Discharge of desalination plant concentrate through a new ocean outfall is widely practiced worldwide and is very popular for projects of all sizes. Over 90% of the large seawater desalination plants in operation dispose of their concentrates through a new ocean outfall specifically designed and built for that purpose. Ocean discharge is typically completed using one of the following methods:

- Direct discharge through new ocean outfall
- Discharge through existing wastewater treatment plant outfall
- Discharge through existing power plant outfall

2.8.3.1.1 Discharge Through a New Ocean Outfall

The purpose of an ocean outfall is to dispose of the plant concentrate in an environmentally safe manner. This primarily means minimization of the size of the zone of discharge in which the salinity is elevated outside of the typical range of tolerance of the aquatic organisms inhabiting the discharge area. The two key options available to accelerate concentrate mixing from an ocean outfall discharge are to rely on the naturally occurring mixing capacity of the tidal (surf) zone or to discharge the concentrate beyond the tidal zone and to install diffusers at the end of the discharge outfall in order to improve mixing.

Although the tidal zone carries a significant amount of turbulent energy and usually provides much better mixing than the end-of-pipe type of diffuser outfall system, this zone has a limited capacity to transport the saline discharge load to the open ocean. If the mass of the saline discharge exceeds the threshold of the tidal zone's salinity load transport capacity, the excess salinity would begin to accumulate in the tidal zone and could ultimately result in a long-term salinity increment in this zone beyond the level of tolerance of the aquatic life. Therefore, the tidal zone is usually a suitable location for discharge only when it has adequate capacity to receive, mix, and transport the salinity discharge to the open ocean. This salinity threshold mixing/transport capacity of the tidal zone can be determined using hydrodynamic modeling. If the TDS discharge load is lower than the tidal zone threshold mixing/transport capacity, then concentrate disposal to this zone is preferable and much more cost-effective than use of a long open outfall equipped with a diffuser system.

For very small plants (i.e., plants with production capacity of 400 m^3/day or less), the ocean outfall is usually constructed as an open-ended pipe that extends up to 100 m into the tidal (surf) zone of the ocean. These discharges usually rely on the mixing turbulence of the tidal zone to dissipate the concentrate and to quickly bring the discharge salinity of the small volume to ambient conditions. Ocean outfalls for large seawater desalination plants typically extend beyond the tidal zone. Large ocean outfalls are equipped with diffusers in order to provide the mixing necessary to prevent the heavy saline discharge plume from accumulating at the ocean bottom in the immediate vicinity of the discharge. The length, size, and configuration of the outfall and diffuser structure for a large desalination plant are typically determined based on hydrodynamic modeling for the site-specific conditions of the discharge location.

2.8.3.1.2 Discharge Through an Existing Wastewater Treatment Plant Outfall

The key feature of this combined discharge method is the benefit of accelerated mixing that stems from blending the heavier high-salinity concentrate with the lighter low-salinity wastewater discharge. Depending on the volume of the concentrate and how well the two waste streams are mixed prior to the point of discharge, the blending may allow the size of the wastewater discharge plume to be reduced and dilute some of its constituents. Co-discharge with the lighter-than-seawater wastewater effluent would also accelerate the dissipation of the saline plume by floating this plume upward and expanding the volume of the ocean water with which it mixes.

Direct discharge through an existing wastewater treatment plant outfall has found a limited application to date, especially for medium-sized and large seawater desalination plants. Key considerations are the availability and cost of wastewater outfall capacity and the potential for whole effluent toxicity of the blended discharge that may result from ion imbalance of the blend of the two waste streams (Mickley 2006). For this concentrate disposal option to be feasible, there must be an existing wastewater treatment plant in the vicinity of the desalination plant, and this plant must have available extra outfall discharge capacity. The fees associated with the use of the wastewater treatment plant outfall must be acceptable, and the wastewater treatment plant utility that would allow the use of their outfall for concentrate discharge must be comfortable with the arrangement of handling and separation of liability for environmental impacts of the blended discharge between the owner of the desalination plant and the owner of the wastewater treatment plant. This beneficial combination of conditions is usually not easy to find, especially for discharging large concentrate volumes.

Bioassay tests completed on blends of desalination plant concentrate and wastewater effluent from the El Estero wastewater treatment plant in Santa Barbara, California (Bay and Greenstein 1992/1993), indicate that this blend can exhibit toxicity to fertilized sea urchin (*Strongylocentrotus purpuratus*) eggs. Parallel tests on desalination plant concentrate diluted to similar TDS concentrations with seawater rather than wastewater effluent did not show such toxic effects on sea urchins. Long-term exposure of red sea urchins to the blend of concentrate and ambient seawater confirms the fact that sea urchins can survive elevated salinity conditions when the discharge is not mixed with wastewater. Thus, the wastewater component is the critical issue.

These findings clearly indicate that blending of wastewater effluent with desalination plant concentrate may have negative effects on some aquatic species, and they need careful consideration when analyzing the advantages and disadvantages of using existing wastewater treatment plant outfalls for concentrate discharge. The aquatic organisms that are recommended for testing of the toxicity of the blend of wastewater effluent and desalination plant concentrate are the echinoderms (Phylum Echinodermata), which include species such as the urchins, starfish, sand dollars, and serpent stars. The echinoderms are the aquatic organisms most sensitive to exposure to a blend of wastewater and concentrate, because they are the only major marine taxa that do not extend into fresh water.

The most likely factor causing toxic effects in the sensitive marine species is the difference in ratios between the major ions (calcium, magnesium, sodium, chloride, and sulfate) and TDS that occurs in the wastewater effluent–concentrate blend compared with the concentrate–seawater blend and the ambient ocean water. As the RO membranes reject key seawater ions except for some complex ions such as borate at approximately the same level, the ratios between the concentrations of those ions and the TDS in the concentrate are approximately the same as these ratios in the ambient seawater. Therefore, marine organisms are not exposed to conditions of ion ratio imbalance if this concentrate is directly discharged to the ocean. As wastewater effluent has a freshwater origin and fresh water often has very different ratios of key ions to TDS, blending this effluent with seawater concentrate may yield a discharge that has significantly different ionic ratios from those of the ambient seawater.

2.8.3.1.3 Discharge Through an Existing Power Plant Outfall (Collocation)

The key feature of the collocation concept is the direct connection of the desalination plant intake and discharge facilities to the discharge outfall of an adjacently located coastal power generation plant (Voutchkov 2004b). This allows use of the power plant cooling water both as source water for the desalination plant and as a blending water to reduce the salinity of the desalination plant concentrate prior to the discharge to the ocean.

For collocation to be cost-effective and feasible to implement, the power plant cooling water discharge flow must be larger than the desalination plant flow, and the power plant outfall configuration must be adequate to avoid entrainment and recirculation of concentrate into the desalination plant intake. Preferably, the length of the power plant outfall downstream of the point of connection of the desalination plant discharge should be adequate to achieve complete mixing prior to the point of entrance into the sea. Special consideration must be given to the effect of the power plant operations on the cooling water quality, as this discharge is used as source water for the desalination plant. For example, if the power plant discharge contains levels of copper, nickel, or iron significantly higher than those of the ambient seawater, this power plant discharge may be not be suitable for collocation of a membrane plant, because these metals may cause irreversible fouling of the membrane elements.

Collocation yields four key benefits:

1. The construction of separate desalination plant intake and outfall structures is avoided, thereby reducing the overall cost of desalinated water.
2. The salinity of the desalination plant discharge is reduced as a result of the mixing and dilution of the membrane concentrate with the power plant discharge, which has ambient seawater salinity.

3. Because a portion of the discharge water is converted into drinking water, the power plant thermal discharge load is decreased, which in turn lessens the negative effect of the power plant thermal plume on the aquatic environment.
4. Blending of the desalination plant and the power plant discharges results in accelerated dissipation of both the salinity and the thermal discharges.

Avoiding the cost of construction of a separate ocean outfall would result in a measurable reduction of plant construction expenditures. In addition, the length and configuration of the desalination plant concentrate discharge outfall are closely related to the discharge salinity. Usually, the lower the discharge salinity, the shorter the outfall and the less sophisticated the discharge diffuser configuration needed to achieve environmentally safe concentrate discharge. Blending the desalination plant concentrate with the lower-salinity power plant cooling water often allows the reduction of the overall salinity of the ocean discharge within the range of natural variability of the seawater at the end of the discharge pipe, thereby completely alleviating the need for complex and costly discharge diffuser structures. In addition, the power plant thermal discharge is lighter than the ambient ocean water because of its elevated temperature; thus, it tends to float on the ocean surface. The heavier saline discharge from the desalination plant draws the lighter cooling water downward and thereby engages the entire depth of the ocean water column into the heat and salinity dissipation process. This phenomenon results in shortening the time for dissipation of both discharges, and the area of their impact is reduced.

2.8.3.2 Concentrate Surface Water Discharge Issues and Considerations

The key challenges associated with selecting the most appropriate location for the desalination plant's ocean outfall discharge are finding an area devoid of endangered species and stressed marine habitats, identifying a location with strong ocean currents that allow quick and effective dissipation of the concentrate discharge, avoiding areas with busy naval vessel traffic that could damage the outfall facility and change mixing patterns, and finding a discharge location in relatively shallow waters that at the same time is close to the shoreline in order to minimize outfall construction expenditures. Key environmental issues and considerations associated with concentrate disposal to surface waters include the following:

- Salinity increase beyond the tolerance thresholds of the species in the area of discharge
- Concentration of metals and radioactive ions to harmful levels
- Concentration and discharge of nutrients that trigger changes in marine flora and fauna in the area of discharge
- Compatibility between the composition of the desalination plant concentrate and receiving waters (ion imbalance-driven toxicity)
- Elevated temperature from thermal desalination processes
- Disturbance of bottom marine flora and fauna during outfall installation

Key issues associated with the feasibility of disposal of seawater desalination plant concentrate to the ocean under any of the discharge methods described earlier include

- Evaluation of discharge dispersion and recirculation of the discharge plume to the plant intake
- Establishment of the marine organism salinity tolerance threshold for the site-specific conditions of the discharge location and outfall configuration
- Evaluation of the potential for whole effluent toxicity of the discharge
- Assessment of whether the discharge water quality meets the numeric and qualitative effluent water quality standards applicable to the point of discharge

2.8.3.2.1 Evaluation of Concentrate Dispersion

The main purpose of the evaluation of the concentrate dispersion from the point of discharge is to establish the size of the zone of initial dilution (ZID) required to dissipate the discharge salinity plume to ambient seawater TDS levels and to determine the TDS concentrations at the surface, at the mid-level of the water column and at the ocean bottom in the ZID. The TDS concentration fields at these three levels are then compared with the salinity tolerance of the marine organisms inhabiting the surface (mostly plankton), those living in the water column (predominantly invertebrates), and the bottom dwellers in order to determine the impact of the concentrate salinity discharge on these organisms.

The discharge salinity field in the ZID and the ZID boundaries is established using hydrodynamic modeling. This modeling allows the determination of the most suitable location, design configuration, and size of the ocean outfall and diffusers if a new outfall is needed or assessment of the feasibility of using existing wastewater or power plant outfall facilities. The model selected for determining the boundaries of the desalination plant discharge should be used to depict the concentrate plume dissipation under a variety of outfall and diffuser configurations and operational conditions. Evaluation of concentrate dispersion and recirculation for large seawater desalination plants usually requires sophisticated plume analysis and is completed using various computational fluid dynamics software packages tailor-made for a given application.

2.8.3.2.2 Aquatic Life Salinity Tolerance Threshold

Many marine organisms are naturally adapted to changes in seawater salinity. These changes occur seasonally and are mostly driven by the evaporation rate from the ocean surface, by rain/snow deposition and runoff events, and by surface water discharges. The natural range of seawater salinity fluctuations could be determined based on information from sampling stations located in the vicinity of the discharge and operated by national, state, or local agencies and research centers responsible for ocean water quality monitoring. Typically, the range of natural salinity fluctuation is at least ±10% of the average annual ambient seawater salinity concentration. The "10% increment above ambient ocean salinity" threshold is a conservative measure of aquatic life tolerance to elevated salinity. The actual salinity tolerance of most marine organisms is usually significantly higher than this level and usually exceeds 40%.

2.8.3.2.3 *Whole Effluent Toxicity Evaluation*

Whole effluent toxicity testing is an important element of the comprehensive evaluation of the effect of the concentrate discharge on aquatic life. Completion of both acute and chronic toxicity testing is recommended for the salinity levels that may occur under a worst-case combination of conditions in the discharge (Voutchkov 2006). Use of at least one species endogenous to the targeted discharge is desirable. In the case of concentrate discharge through an existing wastewater treatment plant outfall, it is recommended that at least one echinoderm species (i.e., urchin, starfish, sand dollar, or serpent star) be tested for a worst-case scenario blend of concentrate and wastewater effluent (typically, maximum wastewater effluent flow discharge combined with average concentrate flow).

2.8.3.2.4 *Compliance with Numeric Effluent Discharge Water Quality Standards*

The key parameters that should be given attention regarding concentrate compliance with the numeric effluent discharge water quality standards are TDS, metals, turbidity, and radionuclides. At present, most countries do not have numeric standards for total TDS discharges.

Because metal content in ocean water is naturally low, compliance with the numeric standards for toxic metals usually does not present a challenge. However, concentrate co-discharge with wastewater treatment plant effluent may occasionally present a concern, because wastewater plant effluent contains metal concentrations that may be higher than those in the ambient ocean water. Similar attention to the metal levels in the combined discharge should be given to co-disposal of power plant cooling water and concentrate, especially if the power plant equipment leaches metals such as copper and nickel, which may then be concentrated in the desalination plant discharge. If the desalination plant has a pretreatment system that uses a coagulant (e.g., ferric sulfate or ferric chloride), the waste discharges from the source water pretreatment may contain elevated concentrations of iron and turbidity that must be accounted for when assessing their total discharge concentrations.

Radionuclide levels in the ocean water often exceed the effluent water quality standards, and the RO system concentrate is likely to contain elevated gross alpha radioactivity. This condition is not unusual for both Pacific Ocean and Atlantic Ocean water and must be well documented with adequate water quality sampling in order to avoid potential permitting challenges.

One important challenge with all concentrate water quality analyses is that most of the laboratory analysis guidelines worldwide are developed for testing fresh water rather than seawater or high-salinity concentrate. The elevated salt content of the concentrate samples could interfere with the standard analytical procedures and often produces erroneous results. Therefore, concentrate analysis must be completed by an analytical laboratory experienced with and properly equipped for seawater analysis. The same recommendation applies for the laboratory retained to complete the whole effluent toxicity testing and source water quality characterization using techniques designed for saline water. This topic is addressed in Chapter 5.

2.8.4 Concentrate Discharge to Sanitary Sewer

Discharge to the nearby wastewater collection system is one of the most widely used methods for disposal of concentrate from brackish water desalination plants (see Table 2.5). This concentrate discharge method is suitable only for disposal of concentrate from very small brackish water and seawater desalination plants into large-capacity wastewater treatment facilities, mainly because of the potential negative effects of the concentrate's high TDS content on the wastewater treatment plant operations. Discharging concentrate to the sanitary sewer is regulated by the requirements applicable to industrial discharges and the applicable discharge regulations of the utility/municipality that is responsible for wastewater collection system management.

The feasibility of this disposal method is limited by the hydraulic capacity of the wastewater collection system and by the treatment capacity of the wastewater treatment plant receiving the discharge. Typically, the wastewater treatment plant's biological treatment process is inhibited by high salinity when the plant influent TDS concentration exceeds 3,000 mg/L. Therefore, before directing desalination plant concentrate to the sanitary sewer, the increase in the wastewater treatment plant influent salinity must be assessed, and its effect on the plant's biological treatment system has to be investigated. Considering that wastewater treatment plant influent TDS may be up to 1,000 mg/L in many facilities located along the ocean coast and that the seawater desalination plant concentrate TDS is likely to be 65,000 mg/L or higher, the capacity of the wastewater treatment plant must be at least 30–35 times higher than the daily volume of concentrate discharge in order to maintain the wastewater plant influent TDS concentration below 3,000 mg/L. For example, a 38,000 m^3/day wastewater treatment plant would likely not be able to accept more than 1,000 m^3/day of concentrate (i.e., serve a seawater desalination plant of capacity greater than 1,000 m^3/day).

If the effluent from the wastewater treatment plant is designated for water reuse, the amount of concentrate that can be accepted by the wastewater treatment plant is limited not only by the concentrate salinity but also by the content of sodium, chloride, boron, and bromide in the blend. All of these could have a profound adverse effect on the reclaimed water quality, especially if the effluent is used for irrigation. Treatment processes in a typical municipal wastewater treatment plant, such as sedimentation, activated sludge treatment, and sand filtration, do not remove a significant amount of these concentrate constituents.

Many crops and plants cannot tolerate irrigation water that contains over 1,000 mg TDS/L. However, TDS is not the only parameter of concern in terms of irrigation water quality. Boron or borate levels in the effluent could also limit agricultural reuse, especially in areas of very low precipitation because borates are herbicides. Chloride and sodium may also have measurable effects on the irrigated plants. Most plants cannot tolerate chloride levels above 250 mg/L. The typical wastewater plant effluent has chloride levels of less than 150 mg/L, whereas the seawater treatment plant concentrate would have chloride concentrations of 50,000 mg/L or more. Using the chloride levels indicated earlier, a 38,000 m^3/day wastewater treatment plant could not accept more than 75 m^3/day of concentrate if the plant's effluent would be used for irrigation. This limitation could be even more stringent if the wastewater effluent were to be used for irrigation of salinity-sensitive ornamental plants. For concentrate discharge from brackish plants, the issues discussed earlier are usually of lesser significance, because the salinity of the concentrate in most cases is not higher than the wastewater influent TDS threshold of 3,000 mg/L. Therefore, direct concentrate discharge to a sanitary sewer is most widely used for brackish water desalination plants and rarely practiced for seawater desalination applications.

2.8.5 Concentrate Deep Well Injection

This disposal method involves the injection of desalination plant concentrate into an acceptable confined deep underground aquifer below a freshwater aquifer using a system of disposal wells. The deep well injection concentrate disposal system also includes a set of monitoring wells to confirm that the concentrate is not migrating to the adjacent aquifers. A variation of this disposal alternative is the injection of concentrate into existing oil and gas fields to aid field recovery. Deep well injection is frequently used for concentrate disposal from all sizes of brackish water desalination plants. Beach well disposal is an alternative concentrate disposal practice. Compared with deep well injection, beach well disposal consists of concentrate discharge into a relatively shallow unconfined coastal aquifer that ultimately conveys this discharge into the open ocean through the ocean bottom. Beach wells are used for small- and medium-sized seawater desalination plants and are not discussed further due to the practice's limited application and success record.

Key issues and considerations associated with deep well injection include the following:

- Limited to site-specific conditions of confined aquifers of large storage capacity that have good soil transmissivity (Schwartz 2000).
- Not feasible for areas of elevated seismic activity or near geological faults that can provide a direct hydraulic connection between the discharge aquifer and a water supply aquifer.
- Potential for contamination of groundwater with concentrated pollutants if the discharge aquifer is not adequately separated from the water supply aquifer in the area of discharge.
- Potential for leakage from the wells.
- Potential scaling and decrease of well discharge capacity over time.
- A backup concentrate disposal method is required for periods of time when the injection wells are tested and maintained.
- High well construction and monitoring costs.

2.8.6 Evaporation Ponds

This method is based on natural solar evaporation of the concentrate in human-made lined earthen ponds or other basins. Evaporation ponds are zero-discharge technology. Holding ponds are needed for concentrate storage, while the evaporation pond reaches the high salinity needed for normal pond operations. Use of evaporation ponds is limited by the following factors:

- Suitable only for disposal of concentrate from small plants in arid areas with low land costs.
- Significant land requirements.
- Require leveled land area.
- Climate dependence.
- Evaporation rate decreases as solids and salinity levels in the ponds increase.
- If the evaporation ponds are not lined, a portion of the concentrate may percolate to the freshwater aquifer beneath the pond.
- Salts accumulated at the bottom of the ponds are ultimately disposed of in a suitable landfill.

2.8.7 Spray Irrigation

This disposal technology uses concentrates for irrigation of salinity-tolerant crops or ornamental plants (lawns, parks, golf courses, etc.). The key issues and constraints associated with spray irrigation are

- Seasonal nature.
- Restricted to small desalination plants.
- A backup disposal alternative is required when crop irrigation is not needed.
- Feasibility determined by climate, land availability, irrigation demand, and salinity tolerance of the irrigated plants.
- Very limited types of crops and ornamental plants that can be grown on high-salinity water.
- Possible negative impact on groundwater aquifer beneath the irrigated area; use of this method may cause significant concerns if the concentrate contains arsenic, nitrates, metals, or other regulated contaminants.

2.8.8 Zero Liquid Discharge

Zero liquid discharge technologies, such as brine concentrators, crystallizers, and dryers, convert concentrate to highly purified water and solid dry product suitable for landfill disposal or perhaps recovery of useful salts.

2.8.8.1 Concentrators (Thermal Evaporators)

Concentrators are single-effect thermal evaporator systems in which the vapor produced from boiling concentrate is pressurized by a vapor compressor. The compressed vapor is then recirculated for more vapor production from the concentrate. The concentrators are typically used for water reuse applications. Usually, the concentrator technology allows evaporation of 90%–98% of the concentrate and produces low-salinity fresh water. The concentrated stream can be further dewatered and disposed of in a landfill as a solid waste. Ultimately, the concentrated salt product could be designated for commercial applications. Vapor compression-driven concentrators are very energy efficient: they use approximately 10 times less energy than single-effect steam-driven evaporators. Typically, energy for the concentrators is supplied by a mechanical vacuum compression system.

2.8.8.2 Crystallizers

Crystallizers are used to extract highly soluble salts from concentrate. The crystallization vessels are vertical units operated using steam supplied by a boiler or heat provided by vacuum compressors for evaporation. Concentrate is fed to the crystallizer vessel, passed through a shell-and-tube heat exchanger, and heated by vapor introduced by the vacuum compressor. The low-salinity water separated from the concentrate is collected as distillate at the condenser. The heated concentrate then enters the crystallizer, where it is rotated in a vortex. Concentrate crystals are formed in the vessel, and the crystalline mineral mass is fed to a centrifuge or a filter press to be dewatered to a solid state. The mineral cake removed from the concentrate contains 85% solids and is the only waste stream produced by the crystallizer.

The energy cost for concentrate evaporation and crystallization is high (260–660 kWh/10 m^3), and the equipment costs are usually several times greater than the capital investment needed for the other concentrate disposal alternatives. Because of the high capital and operation and maintenance costs, the zero-discharge technologies are not practical unless no other concentrate management alternatives are available. Usually, zero-discharge concentrate management systems are justifiable for inland brackish water desalination plants where site-specific constraints limit the use of natural evaporation or wastewater treatment plant disposal. High-recovery desalination systems preceding the evaporators reduce energy and capital costs.

2.8.9 Regional Concentrate Management

Regional concentrate management includes two alternative approaches that can be used separately or co-implemented at the same facility:

1. Regional collection and centralized disposal of seawater concentrate through one location applying one or more of the methods described in the previous sections.
2. Use of concentrate from brackish water desalination plants as source water to a seawater desalination plant.

The first approach takes advantage of site-specific beneficial conditions for disposal that may not be available at other locations and of the economies of scale of constructing larger concentrate disposal facilities. The second approach takes advantage of the fact that concentrate from brackish water plants is of significantly lower salinity than seawater and, when blended with the ocean water fed to a seawater desalination plant, will reduce the overall plant salinity. The use of concentrate from a brackish water desalination plant as feedwater to a seawater desalination plant is mutually beneficial for both plants. Usually, inland brackish water desalination plant capacity is limited by the lack of suitable discharge locations for the plant concentrate. If the seawater desalination plant can accept the brackish water desalination plant concentrate and process it, the brackish water desalination plant capacity could be increased beyond the threshold driven by brine discharge limitations, and the desalination plant source salinity could be reduced at the same time. Whereas the seawater TDS concentration is usually in a range of 30,000–40,000 mg/L, the TDS of the concentrate from brackish water desalination plants is typically several times lower (i.e., typically 2,000–15,000 mg/L). Therefore, when blended with the source seawater, it would reduce the overall desalination plant feedwater salinity, which would have a positive effect on the desalination plant power use, recovery factor, and cost of water. The key considerations associated with regional concentrate disposal stem from the location of a large volume of discharge in a single point. This may amplify the negative effect of disposal on the marine organisms in the area of discharge.

2.8.10 Technologies for Beneficial Use of Concentrate

Concentrate from seawater desalination plants contains large quantities of minerals that may have commercial value when extracted. The most valuable minerals are magnesium, calcium, sodium chloride, and bromine. Magnesium compounds in seawater have agricultural, nutritional, chemical, construction, and industrial applications. Calcium sulfate (gypsum) is used as a construction material for wallboard, plaster, building cement, and road building and repair. Sodium chloride can be used for the production of chlorine and caustic soda, highway deicing, and food products. Technologies for the beneficial recovery of minerals from concentrate can be used for the management of concentrate from both inland brackish water desalination plants and coastal seawater desalination plants. These technologies have the potential to decrease the volume and cost of transporting concentrate as well. Specific emerging technologies for beneficial reuse of desalination plant concentrate are discussed in the following text.

2.8.10.1 Salt Solidification and Recovery

The existing salt recovery technologies extract salts by fractional crystallization or precipitation. Crystallization of a given salt can be achieved by concentrate evaporation or temperature control. Fractional precipitation is attained by adding a precipitating chemical agent to selectively remove a target mineral from the concentrate solution. For example, there is a commercially available technology for extraction of magnesium and calcium salts from concentrate and for production of structural materials from these salts.

2.8.10.2 Disposal to Brackish or Saltwater Wetlands

This method is site specific and suitable for conditions where the concentrate quality is compatible with the native flora and fauna of the saltwater marsh or wetland. Usually, the wetlands or marshes that would be used for concentrate discharge are hydraulically interconnected with the ocean or a brackish water body; therefore, this is an indirect method for concentrate disposal to surface waters. Wetland vegetation may assimilate some of the nitrate and selenium in the concentrate, providing effective reduction of these contaminants.

2.8.10.3 Concentrate Use as Cooling Water

Use of concentrate for power plant cooling is typically practiced for small power plants with limited cooling needs and cooling towers that are made of materials suitable to withstand the highly corrosive concentrate. A key concern is the high scaling potential of the concentrate.

2.8.10.4 Other Beneficial Uses

Small volumes of concentrate have been used occasionally for dust suppression, roadbed stabilization, soil remediation, and deicing. Such applications are very site specific and can be used only as a supplemental concentrate disposal alternative.

2.9 MANAGEMENT OF RESIDUALS GENERATED AT DESALINATION PLANTS

2.9.1 Pretreatment Process Residuals

Table 2.7 lists residuals that may be produced in the pretreatment process when applying membrane desalination. The amount of residuals produced is primarily a

TABLE 2.7
Residuals from Membrane Desalination Processes

Residual	Source or Cause	Application
Backwash solids/sludge	Suspended solids in the feedwater	Seawater, brackish water, wastewater
Backwash water	From removal of suspended solids in the feedwater	Seawater, brackish surface water, wastewater
Cleaning solutions (see Table 2.3)	Cleaning of filtration membranes (MF/UF) and process membranes (RO, NF, ED)	Seawater, brackish water, wastewater
Concentrate or blowdown (if cross-flow)	Filtration membranes operating in a cross-flow mode	Seawater, brackish water, wastewater
Spent media (sand, anthracite, and/or garnet)	From the removal of suspended solids in the feedwater	Seawater, brackish water, wastewater
Cartridge filters—polypropylene	Final fine filtration prior to RO, periodic replacement	Most membrane desalination processes, except those using MF/UF filtration membranes
MF/UF membranes—polymeric material (polypropylene, polysulfone, polyvinylidene fluoride, cellulose acetate)	Membrane replacement for MF/UF systems	Seawater, brackish water, wastewater
RO membranes (polyamide thin-film composite, cellulose acetate)	Membrane replacements	Seawater, brackish water, wastewater

Source: Courtesy of N. Voutchkov.

function of the feedwater quality relative to the constituents that must be removed prior to the membrane desalination process. Surface water, seawater, and wastewater for reuse all have significant levels of suspended solids, for example. These solids must be removed prior to treatment with reverse osmosis, either in a backwash stream or as sludge. Other than the concentrate stream, these solids create the most significant residual stream from a desalination plant.

2.9.2 Management of Spent Pretreatment Filter Backwash Water

Spent filter backwash water is a waste stream produced by the membrane plant's pretreatment filtration system. Depending on the type of pretreatment system used (granular or membrane filters), the spent filter backwash water will vary in quantity and quality. In general, the membrane pretreatment systems produce 1.5–2 times larger volumes of spent filter backwash water than do the granular media filters. However, compared with MF or UF membrane pretreatment filters, granular media filters typically require larger dosages of coagulant for pretreatment. Depending on the source water quality, membrane pretreatment may allow pretreatment to be successful without coagulant addition. Spent pretreatment filter backwash water may include filter aids and coagulants.

Spent filter backwash water can be handled using one of the following methods. Discharge to a surface water body along with plant concentrate without treatment is one of the most widely practiced disposal methods. This is typically the lowest-cost disposal method because it does not involve any treatment prior to disposal. It is suitable for discharge to large water bodies with good flushing, such as open oceans or large rivers. On-site treatment prior to surface water discharge and recycle upstream of the filtration system are possible options. The filter backwash water must be treated at the membrane treatment plant when its direct discharge does not meet surface body water quality requirements or discharged to deep injection wells if it is not suitable for direct disposal. Typically, the most widely used granular media backwash treatment method is gravity settling in conventional or lamella plate sedimentation tanks. Spent wash water from membrane pretreatment systems is usually treated in separate MF or UF membrane modules. Filter backwash sedimentation tanks are often designed for a retention time of 3–4 h and allow the removal of more than 90% of the backwash solids. The settled filter backwash water can be either disposed of with the membrane plant concentrate or recycled at the head of the pretreatment filtration system. It may be more cost-effective to recycle and reuse the settled filter backwash water rather than to dispose of it with the concentrate. However, blending and disposal with the concentrate may be more beneficial if the concentrate water quality is inferior and cannot be disposed of to a surface water body without prior dilution with a stream of lesser salinity. The solid residuals (sludge) retained in the sedimentation basin are often discharged to the sanitary sewer in liquid form (typically practiced at small- to medium-sized plants) or dewatered on-site in a designated solids-handling facility.

2.9.3 Management of Spent (Used) Membrane Cleaning Solutions

The accumulation of silt or scale on the membranes causes fouling, which reduces membrane performance. Desalination system membranes must be cleaned periodically to remove foulants and extend the membrane's useful life. Typical cleaning frequency of the membranes is two to four times per year. Membrane trains are usually cleaned sequentially. A chemical cleaning solution is circulated through the membrane train for a preset time. After the cleaning solution circulation is completed, the spent cleaning solution is evacuated from the train to a storage tank, and the membranes are flushed with permeate (flush water). The flush water is used to remove all the residual cleaning solution from the RO train in order to prepare the train for normal operation. The flush water is stored separately from the rest of the plant permeate in a flush tank.

The various waste discharge volumes generated during the membrane train cleaning process are as follows:

- Concentrated waste cleaning solution is the actual spent membrane cleaning chemical.
- Flush water residual cleaning solution (first flush) is the first batch of clean product water used to flush the membranes after the recirculation of cleaning solution is discontinued. This first flush contains diluted residual cleaning solution.
- Flush water permeate is the spent cleaning water used for several consecutive membrane flushes after the first flush. This flush water is of low salinity and contains only trace amounts of cleaning solution.
- Flush water concentrate is the flush water removed from the concentrate lines of the membrane system during the flushing process. This water contains very little cleaning chemicals and is of slightly higher salinity than the flushing permeate.

All the membrane-cleaning streams listed earlier are typically conveyed to one washwater tank often named the "scavenger tank" for waste-cleaning solution retention and treatment. This tank must be able to retain the waste-cleaning solution from the simultaneous cleaning of a minimum of two membrane trains. The scavenger tank should be equipped with mixing and pH neutralization systems. The mixing system should be installed at the bottom of the tanks to provide complete mixing of all four cleaning solution streams listed earlier. After mixing with flush water, the concentration of the cleaning solution chemicals will be reduced significantly. The used cleaning solution should be neutralized to a pH compatible with the pH requirements for discharge to the wastewater collection system. At many plants, only the most concentrated first flush is discharged to the wastewater collection system. The rest of the flush water usually has only trace levels of contaminants and is most often suitable for a surface water discharge (i.e., discharge to the ocean or other nearby water body).

2.10 SMALL DESALINATION SYSTEMS

Small desalination systems are facilities that produce 4,000 m^3/day or less of drinking water. In most cases, small desalination systems use treatment technologies similar to those applied for large desalination plants. Small desalination systems may be classified in various categories depending on their technology, mobility, and service. In terms of technology, small desalination systems, similar to large plants, can be divided into two groups: membrane and thermal facilities. In terms of mobility, small plants can be divided into stationary and mobile units. Depending on their service, small plants are classified as single point-of-use and multiple point-of-use systems. All of these systems should be tested against performance standards and should demonstrate their ability to reliably produce safe drinking water under their appropriate use conditions. The specific areas of allocation and technology used for the various small desalination systems are described in the following text.

2.10.1 Small Applications for Thermal Desalination

Mechanical vapor compression (MVC) systems are a special subset of MED evaporators that incorporate a rotating mechanical vapor compressor as the means of heating the process (Figure 2.8). Typically, mechanical vapor compressors are rotated by electric motors or sometimes engines. MVC systems were very common for small applications prior to the commercial availability of RO plants. These plants can produce distilled water from almost any salinity of feedwater, yet they do not require a heat source (and therefore do not need cooling water). Typically, an MVC unit has a recovery of 40%–50% when operating on seawater and can yield much higher recoveries (70%–85%) when using brackish water. MVC systems have found applications in remote locations with robust equipment requirements, such as oil rigs and some marine vessels, and for production of boiler feedwater for power plants from seawater. Other MVC applications include the production of potable water for bottling and various pharmaceutical grades of water. Most MVC plants produce 10–200 m^3/day of distilled water, although a few installations producing up to 2,500 m^3/day have been built.

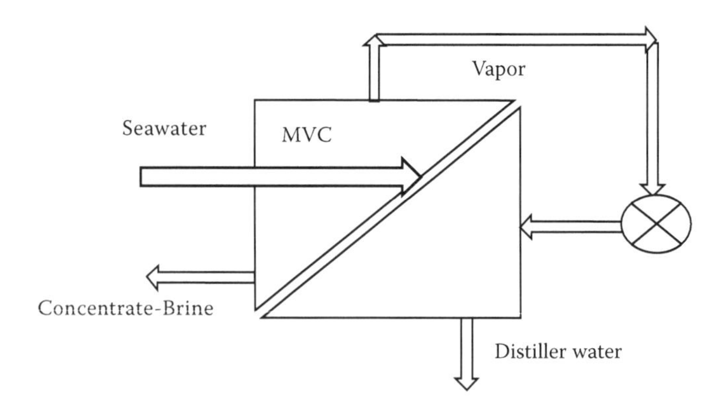

FIGURE 2.8 General schematic of a mechanical vapor compression (MVC) unit.

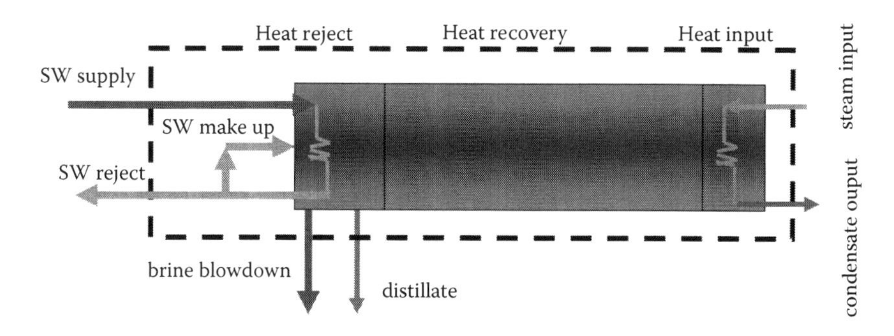

FIGURE 2.9 General schematic of a small distiller unit.

Small distillers (Figure 2.9) are another special subset of MED systems that are usually configured to use hot water from engines or other industrial processes. Typically, the systems have only one or two effects, making them thermodynamically inefficient. However, they are designed for small-volume applications in situations concurrent with large quantities of heat (which generally use once-through cooling whether or not desalination is present). Similar to MVC plants, small stills have found some application in oil rigs, in marine vessels (including cruise ships) and for boiler water production for power plants. These units typically have capacities between 1 and 50 m^3/day and usually have maximum evaporation temperatures of 60°C–80°C.

Small MSF systems are often found in merchant marine, naval, and commercial cruise ships. Steam or hot water from marine diesel engines may provide the heat to drive the process. Typically, these systems are used in applications where there is a small localized demand for water coincident with a relatively large amount of (waste) heat. Small thermal MSF systems produce between 100 and 1,000 m^3 of water per day. These systems are typically not used to produce drinking water and are typically applied for industrial purposes.

The key issues for small thermal systems are the same as for large units. The only exception is that small stills and small MSF systems are typically designed for offshore marine applications, with the result that onshore use may require a disproportionately high cooling water flow.

2.10.2 Small Membrane Desalination Plants

Small membrane desalination plants have found wide use in land-based applications for supplying potable water to hotels and resorts, to small and remote residential communities, and to military installations. This type of plant is also very popular for sea-based mobile applications. Small membrane desalination plants are typically compact seawater or brackish water treatment systems that are configured in one skid and often referred to as *package plants*. For small skid-mounted systems, it is common to have the pretreatment chemical injection points, cartridge filter, feed pump, membranes, and pressure vessels on the skid. For larger systems, the pretreatment components and feed pumps are typically not located on the rack supporting the membranes and the pressure vessels. In contrast to large membrane desalination systems, which are designed to minimize overall life cycle water production costs, small systems are typically configured and designed with features that aim to simplify facility installation and operation. Often, these systems are built and assembled in a manufacturing shop away from their point of application and then shipped in a container to the location of their installation and use. Large desalination plants typically use ~20 cm membrane elements; small-membrane desalination units may be designed to operate with smaller size (~5 or 10 cm) membranes. In addition, these plants often apply a more simplified pretreatment system that may include just a series of 20 μm, 10 μm, and 1 μm cartridge filters, rather than granular or membrane pretreatment filters. Small membrane desalination units often use a number of water quality polishing technologies, such as activated carbon filters, ion exchange systems, and UV disinfection systems.

Although small membrane desalination plants use RO or NF membranes similar to those used by large plants, these systems are often not as sophisticated in terms of automated source and product water quality monitoring and control. Therefore, operator training and skills are very important in order to maintain safe and cost-effective drinking water production. If a skilled operations staff is not readily available, the system owner should request the system supplier to design the system with automated shutdown provisions that trigger discontinuation of system operations when the integrity of the pretreatment system or membrane system equipment is breached or when unusually high values of key parameters such as turbidity and conductivity are reached. A series of disinfection and pathogen control devices such as carbon filters followed by 1 μm cartridge filters followed by UV irradiation or chemical disinfection is recommended, rather than reliance on a single technology for disinfection, especially if the system is not run by a skilled operator or is unmanned for the majority of the time.

2.10.3 Small Stationary Desalination Plants

Small stationary desalination plants may use both thermal and membrane desalination technologies; however, unless an easy-to-access steam source exists, membrane desalination systems are usually the preferred choice for small stationary applications due to their more compact size and lower overall energy demand. A typical stationary membrane desalination system will have the following key elements located on one skid: connection piping to a water source; booster pump to pretreatment system; sand/solids separator; granular multimedia pressure filters or membrane filters; 1–8 membrane vessels with salt separation membranes; chemical injection systems for coagulant, biocides, sodium bisulfite, and antiscalant; and a disinfection system using UV irradiation or chlorination.

2.10.4 Mobile Desalination Plants for Emergency Water Supply

Mobile desalination plants are easy-to-transport modular water treatment systems that are designed to be deployed in zones of natural calamities (earthquakes, floods, etc.), military operations, and other humanitarian crisis conditions and to operate with practically any water supply source, including waters contaminated with raw wastewater, high solids content, oil and grease, organics, pesticides, and wastewater. The water treatment equipment incorporated in these types of systems is installed in sturdy metal frames (racks) designed to withstand the weather elements and long-distance transportation on trucks and airplanes. The source water pretreatment system typically includes chlorination and dechlorination, strainers, grit separators and centrifuges for solids removal, granular media filtration (including sand filters and activated carbon filters), followed by cartridge filters and an RO membrane system. In some cases, pretreatment may be provided by a UF system instead of a granular media filter. Drinking water disinfection for such systems is provided by chlorination and UV irradiation. These units are usually supplied with electricity by a diesel-powered generator.

Compared with stationary and large desalination plants, the operation of mobile systems requires substantially larger amounts of consumables, especially if the source water is contaminated with oils and is very turbid. The operation of these systems is limited by the energy available. The water quality produced by some of these systems may not always meet drinking water regulations for all parameters, but the systems can be designed to produce water of quality adequate for protection of health.

2.10.5 Marine Vessel (Ship/Boat) Desalination Plants

Most marine vessel installations, including cruise ships, military ships, yachts, and boats, use membrane seawater desalination for freshwater supply. This type of small desalination plant is typically constructed with lightweight epoxy powder-coated aluminum or steel frames and can be provided with explosion proofing and fully automated monitoring and control systems that maintain consistent water quality at all times. Usually, the marine vessel membrane desalination plants have the following elements: source water connection pipe, booster pump, dual-media granular filters, oil/water separator, high-pressure feed pump, high-rejection/high-flux RO membrane elements, corrosion-proof fiberglass pressure vessels, and product water postfiltration skid with activated carbon unit and UV disinfection system. For applications with seawater of potentially high solids content, hydrocyclone separators are added ahead of the granular media filters for longer cartridge filter and RO membrane life.

2.10.6 Offshore Seawater Desalination Facilities

Currently, there are commercially available membrane seawater desalination plants installed on large ships that are specifically designed and built with the prime purpose of producing fresh water (Sessions and Hawkins 2006). These facilities offer some benefits over the land-based desalination facilities, such as avoiding the construction of costly source water intake and ocean outfall discharge facilities, reducing the impact of the concentrate discharge on the marine environment, collecting source water from pristine offshore waters, and simplifying the plant permitting process. The intake structure of the vessel is constructed as part of the vessel hull and is equipped with screening facilities. The pretreatment system is a simplified bag filtration facility housed in an HDPE casting. The RO system incorporates high-pressure pumps and membrane racks and is equipped with energy recovery equipment. The produced water is stored on the ship. The marine vessel that is used for this floating water production facility will be connected to a shore terminal via seabed pipeline, which will deliver the potable water to the shore. The actual water production will be completed offshore. During the production of desalinated water, the ship is typically located adjacent to a natural gas platform that will be used to run the desalination plant pumps. These pumps are equipped with dual-fuel gas turbines.

The systems discussed earlier must be suited to the particular application to provide reliable and consistent performance. Marine-based systems must meet the safety standards applicable for the marine industry. Independent performance testing is recommended to confirm conformance with applicable regulations prior to deployment. The system should be equipped with automated freshwater flush at each shutdown and approximately every 7 days to maintain the performance integrity of the membranes.

2.11 SUMMARY AND CONCLUSIONS

Seawater is typically desalinated using two general types of water treatment technologies: thermal evaporation (distillation) and RO membrane separation. Both types of technologies are very reliable and are capable of producing high-quality drinking water of consistent quality and quantity.

2.11.1 Desalination System Intake Facilities

Two general types of intake facilities are used to obtain source water for desalination plants: subsurface intakes (wells, infiltration galleries, etc.) and open intakes. Wells are either vertical or horizontal source water collectors typically located on the seashore, in close proximity to the ocean. These intake facilities are relatively simple to build, and the seawater they collect is naturally pretreated via slow filtration through the subsurface sand/seabed formations in the area of source water extraction. Raw seawater collected using wells is usually of better quality in terms of solids, silt, oil and grease, natural organic contamination, and aquatic microorganisms, compared with open seawater intakes. Beach intakes may sometimes also yield source water of lower salinity. Vertical beach wells are usually less costly than the horizontal wells; however, their productivity is relatively low (typically 400–4,000 m^3/day), and therefore the use of vertical wells for large plans is less favorable. The most widely used type of horizontal collector beach wells is often referred to as Ranney collector wells.

Open ocean intakes for large seawater desalination plants are often complex structures, including intake piping that typically extends several hundred to several thousand meters into the ocean. Source water collected through open intakes usually requires pretreatment prior to RO desalination. The cost and time for construction of a new open ocean intake could be significant and could reach 10%–20% of the overall desalination plant construction cost. Open ocean intakes would result in some entrainment of aquatic organisms compared with beach wells because they take raw seawater directly from the ocean rather than source water prefiltered through the coastal sand formations.

2.11.2 Well Intake Considerations

Key drinking water quality considerations when using well intakes include the following:

- The subsurface conditions and sustainable capacity of the source water aquifer must be commensurate with the volume and quality of the drinking water. Well intakes are usually suitable for desalination plants of capacity lower than 20,000 m^3/day.
- Source water aquifer contamination with human-made pollutants such as NDMA, petroleum products, pesticides, pharmaceuticals, groundwater leakage from industrial facilities, mortuaries, etc., may make the brackish or coastal aquifer unsuitable for production of drinking water or require very elaborate and costly pretreatment and product water conditioning to inactivate these pollutants.
- Elevated contents of iron and manganese in the source water may require the use of elaborate pretreatment technology for their removal, such as greensand filters.
- Elevated levels of hydrogen sulfide in the source water aquifer will have a negative impact on the taste and odor of the drinking water.
- Installation of well intakes in wetland areas is not desirable, because it may cause wetland drainage and significant environmental impacts. In addition, such source water may be of fluctuating quality, which in turn would require more elaborate treatment.
- Intake wells usually collect colder water than open ocean intakes. Desalination of colder water requires more energy and would be more costly.

2.11.3 Open Intake Considerations

- Open intakes are suitable for desalination plants of all sizes.
- Source water from open intakes is typically of higher turbidity, silt, and organic content than well water. Therefore, the use of this type of intake usually requires more elaborate source water pretreatment to produce the same drinking water quality.
- Open intakes of desalination plants should not be located near (within 500 m of) a water and wastewater treatment plant discharge. Water and wastewater plant discharges typically contain elevated concentrations of metals such as iron, copper, nickel, aluminum, and lead, which may cause irreversible fouling of desalination membranes and interrupt drinking water production.
- Open intakes should not be located in large industrial and municipal ports because of the increased potential for source and drinking water contamination with petroleum products, raw wastewater, and waste from chemical spills.

- Seawater open intakes should not be located in the vicinity of freshwater discharges to the ocean, such as river estuaries, coastal marshes, and large storm drain canals, because these discharges may carry large amounts of silt and organic materials that may cause significant biofouling of the membranes and desalination system equipment and may increase the DBP content of the drinking water.
- Algal blooms negatively impact source water quality and may result in elevated organics in the source water and accelerated biofouling of SWRO installations. Red tide algal blooms may result in the release of algal toxins of small molecular weight, such as domoic acid and saxitoxin, which may have an impact on product water quality; however, these are well removed by desalination treatment. In order to minimize the entrance of algal material in the open intake, the source water intake structure should be located at least 5 m under the ocean surface, and the intake should be designed so that the water entrance velocity is less than 0.2 m/s. The low entrance velocity would also minimize the entrainment of marine organisms with the collected source water.
- The intake or discharge of an existing power plant should be used to collect source water for the desalination plant, if feasible, to minimize the need for construction of new intake and discharge and the associated disturbance of the benthic organisms in the intake area.
- The desalination plant's source water intake should not be located in ocean zones of endangered or rare species to minimize the environmental impact associated with the construction of desalination plants.

2.11.4 Source Water Pretreatment Facilities and Chemicals

The need and type of source water pretreatment depend on the source water quality and the type of technology used: thermal or membrane desalination. In all cases, the chemicals used for pretreatment should be of high quality ("drinking water grade") and should contain very low levels of impurities, such as metals and dust. Use of lower-quality, industrial-grade water chemicals for the production of drinking water should be avoided.

Specific guidelines associated with pretreatment systems for membrane and thermal desalination plants are discussed in the following text.

2.11.4.1 Pretreatment for Membrane Desalination Operations

The following drinking-water quality-related issues should be considered when selecting pretreatment for membrane desalination systems:

- Intermittent rather than continuous source water chlorination is recommended because chlorine addition creates large amounts of DBPs (TTHMs of 500–2,000 mg/L or higher), especially if the source water has high organic levels.

- The pretreatment system effluent should be dechlorinated after chlorination to protect the structural integrity of the SWRO membranes and to produce consistent product water quality.
- Raw seawater should not be chloraminated prior to membrane separation. Chloramination generates both chloramines and bromamines. Whereas chloramines are disinfectants with low oxidation potential and do not present a threat for the SWRO membranes, bromamines have order-of-magnitude-higher oxidation strength, and their presence in the feedwater to the SWRO system can cause rapid loss of membrane integrity and deterioration of drinking water quality. Chlorine or chlorine dioxide are preferred to control excessive biogrowth in the intake, pretreatment, and RO systems.
- Use of ferric salts for source water coagulation prior to filtration may result in reddish discoloring of the plant discharge. The spent filter backwash water pretreatment systems should be treated using coagulants (ferric sulfate, ferric chloride, etc.), and this sidestream should be settled prior to discharge to surface waters in order to avoid discoloration.

2.11.4.2 Pretreatment for Thermal Desalination Systems

- Performance of thermal desalination systems is significantly less sensitive to source water quality than that of membrane desalination plants. Therefore, physical pretreatment of the source water is simpler and is often limited to screening to remove coarse debris in order to prevent equipment erosion by suspended solids.
- Chemical pretreatment of thermal desalination systems is more robust than that of membrane systems and is aimed at minimizing scaling and corrosion of the heat exchanger surfaces and minimizing the effects of oil and grease from the source water.
- The cooling water stream is treated to control fouling using an oxidizing agent or biocide.
- Makeup water is conditioned with scale inhibitors.

2.11.5 Thermal Desalination Processes

- Thermal desalination is the predominant technology for seawater desalination under conditions where power generation and water production are combined. The three key technologies used for thermal desalination are MSF, MED, and VCD.
- MSF is the most widely used thermal desalination technology.
- Thermal desalination plants produce distilled water of very low TDS (2–50 mg/L), boron, sodium, and bromide.
- A potential exists for thermal desalination processes to distill volatile organic compounds, if they are present in the source water, unless they are removed by venting, external pretreatment, or reaeration.

- Thermal desalination results in a cooling water discharge of elevated temperature (8°C–12°C). This cooling water is commonly blended with the brine from the distillation process. The brine's temperature is typically 5°C–25°C warmer than that of the ambient seawater.
- Cooling water from thermal desalination treatment processes may contain trace amounts of corrosion inhibitors and DBPs.

2.11.6 Membrane Desalination Systems

- Membrane desalination is growing more rapidly than thermal processes and is a prime choice for areas of the world where low-cost fuel sources are not readily available and power and water production facilities are separated.
- Membrane desalination is practiced using brackish water, seawater, and highly treated wastewater effluent as source water.
- There are two principal types of membrane desalination: RO and ED. ED-based desalination is widely used for low-salinity water sources (TDS <3,000 mg/L). RO and NF are used for source water of any salinity and type.
- Seawater RO membrane systems provide effective removal of pathogens (bacteria, viruses, *Giardia*, and *Cryptosporidium*) and organics of molecular size of approximately 200 Da or larger (e.g., DBPs, algal toxins).
- ED systems do not provide pathogen reduction.
- RO system performance and efficiency are dependent on source water quality (TDS, temperature, scaling compounds, etc.) and physical and chemical pretreatment.
- The level of pretreatment needed for successful operation is primarily a function of source water quality of solids, oxidizing agents, oil and grease, and temperature.
- Membrane material integrity is dependent on various factors, including source water pH, temperature, organic content, concentration of oxidants and oil and grease in the water, and solids content.

2.11.7 Posttreatment

- Posttreatment of the product water from the desalination processes includes stabilization by addition of carbonate alkalinity, corrosion inhibition, remineralization by blending with source water, disinfection, and enhanced removal of specific chemicals (e.g., boron, silica, NDMA).
- Stabilization by addition of calcium carbonate alkalinity is the most widely used approach for corrosion control of metallic pipelines and distribution systems.
- Corrosion inhibition is the most popular posttreatment method for plastic pipelines and distribution systems.
- Sodium hypochlorite and chlorine gas are most widely used for disinfection of desalinated water.

- Disinfection to ensure public health protection is of greater concern than by-product formation, although techniques are available to achieve both adequate disinfection and by-product control.
- Use of chloramines instead of chlorine for disinfection is more advantageous when product water must be conveyed over long distances (over 100 km) or stored for long periods of time (several days) owing to the significantly lower decay rate of chloramines compared with free chlorine.
- Use of ozone as a disinfectant for desalinated water has the potential of forming brominated DBPs, including bromate, due to the presence of bromide.
- Blending of desalinated water for remineralization is suitable with brackish water and only up to about 1% with seawater. The raw water used for blending should be pretreated for chemical and microbial control prior to mixing with the desalinated water.

2.11.8 Concentrate Management

- Concentrate (brine) generated during the desalination process should be disposed of in an environmentally safe manner.
- The concentration of minerals in the brine is usually 2–10 times higher than that of the source water and is a function of the TDS content of the source water and the plant recovery.
- The most widely used method of disposal of concentrate from seawater desalination plants is discharge to a nearby surface water body via outfalls.
- Co-discharge of desalination plant concentrate and cooling water from power plants is advantageous where possible because it accelerates dissipation of both the high-salinity plume from the desalination plant and the thermal plume from the power plant.
- Co-discharge with wastewater treatment plant effluent may cause potential toxicity problems with certain marine organisms (e.g., red sea urchins, stars, sand dollars) that may not adapt easily to the blend.
- Concentrate discharge to sanitary sewers affects wastewater treatment plant performance and the suitability of the treatment plant effluent for reuse.
- Technologies for the beneficial reuse of concentrate may offer a number of benefits, if viable.

2.11.9 Management of Residuals from Pretreatment Systems

- Use of ferric salts for source water coagulation prior to filtration may result in reddish discoloring of the plant discharge (see Section 2.11.4.1).
- Use of MF and UF pretreatment systems typically results in the production of 50%–80% less residual solids than that of conventional granular media filtration. However, UF and MF systems generate a 3%–5% larger volume of waste backwash water than do granular media filters.

- Currently, spent backwash water from pretreatment filter systems is typically discharged to the ocean without treatment. However, if the spent backwash water is planned to be discharged to areas with impaired marine life, then spent filter backwash treatment by sedimentation and sludge dewatering is recommended. It is recommended that the dewatered solids removed from the source water be disposed of to a lined sanitary landfill.
- Spent chemical cleaning solutions from UF, MF, and SWRO membrane treatment should not be discharged to the ocean but should be disposed of to the sanitary sewer instead. Membrane flush water generated after the disposal of the membrane cleaning chemicals is typically safe for disposal to the ocean.

2.11.10 Small Desalination Systems

- Small membrane desalination systems have found wide application for the production of drinking water on cruise ships, military marine, vessels, and boats.
- Easy-to-transport package desalination plants are suitable to provide emergency water supplies to areas impacted by natural disasters.

2.12 RESEARCH RECOMMENDATIONS

- Develop methods for the analysis of metals, suspended solids, and organic compounds in brackish water and seawater to be able to better characterize source water quality. Existing standard analytical methods are developed for the analysis of low-salinity water, and they are affected by the high level of dissolved solids in seawater and brackish water.
- Document the rejection of frequently encountered algal toxins (saxitoxin, domoic acid, okadaic acid, etc.) from seawater by RO membranes and the removal of these compounds by thermal desalination systems.
- Document the rejection of frequently encountered human-made organic contaminants (pharmaceuticals, cosmetic products, NDMA, etc.) by seawater and brackish water membranes, and determine the removal of these compounds by thermal desalination systems.
- Determine strategies for reliable disinfection and corrosion inhibition of desalinated water conveyed over long distances in warm climates.
- Develop reliable methods for online assessment of the integrity of SWRO membranes.
- Document the log removal of key pathogens (total coliforms, *Giardia*, *Cryptosporidium*, bacteria, including *Legionella*, and viruses) by brackish water and seawater desalination membranes and thermal desalination systems.
- Study the use of chlorine dioxide for pretreatment of seawater treated by RO.
- Document the effect of various pretreatment disinfectants on the level and type of DBPs in the desalinated seawater produced by SWRO systems.

BIBLIOGRAPHY AND SUPPORTING DOCUMENTS

USEPA. 2003. *Membrane filtration guidance manual.* Washington: U.S. Environmental Protection Agency, Office of Ground Water and Drinking Water, Standards and Risk Management Division, Technical Support Center.

Wangnick, K. 2004. *IDA worldwide desalination plants inventory report.* Wangnick Consulting for International Desalination Association.

Wisner, B. and J. Adams (eds.). 2003. *Environmental health in emergencies and disasters: A practical guide.* Geneva, World Health Organization. http://www.who.int/water_sanitation_health/emergencies/emergencies2002/en/index.html.

REFERENCES

AWWA. 1996. *Rothberg, Tamburini and Winsor model for corrosion control and process chemistry.* Denver: American Water Works Association.

Bay, S., and D. Greenstein. 1992/1993. Toxic effects of elevated salinity and desalination waste brine. In *SCCWRP 1992–93 annual report.* Costa Mesa: Southern California Coastal Water Research Project, 149–53. ftp://ftp.sccwrp.org/pub/download/DOCUMENTS/AnnualReports/1992_93AnnualReport/ar14.pdf.

Belluati, M., E. Danesi, G. Petrucci et al. 2007. Chlorine dioxide disinfection technology to avoid bromate formation in desalinated seawater in potable waterworks. *Desalination* 203(1–3): 312–8.

City of Carlsbad. 2005. *Environmental impact report for precise development plan and desalination plant project* (EIR SCH No. 2004041081).

City of Huntington Beach. 2005. *Draft recirculated environmental impact report: Seawater desalination project at Huntington Beach.* April.

FAO. 2005. *Proceedings of the FAO Expert Consultation on Water Desalination for Agricultural Applications, 26–27 April 2004.* Rome: Food and Agriculture Organization of the United Nations.

Gordon, G. 2001. Is all chlorine dioxide created equal? *J. Am. Water Works Assoc.* 93(4): 163–74.

Larson, T. 1970. *Corrosion by domestic waters.* Urbana: State of Illinois Department of Registration and Education, Division of the State Water Survey (Illinois State Water Survey Bulletin 56).

McGuire Environmental & Poseidon Resources. 2004. *Disinfection by-product formation in simulated distribution system: Blending desalinated seawater from the Poseidon Resources Inc. pilot facility with local drinking water sources.* Final report. March.

Mickley, M.C. 2006. *Membrane concentrate disposal: Practices and regulation.* 2nd ed. Prepared by Mickley & Associates, Boulder, for the U.S. Department of the Interior, Bureau of Reclamation, Denver (Desalination and Water Purification Research Program Report No. 123). http://www.usbr.gov/pmts/water/media/pdfs/report123.pdf.

Missimer, T.M. 1999. Raw water quality—The critical design factor for brackish water reverse osmosis treatment facilities. *Desalination & Water Reuse* 9, no. 1 (May/June): 41–7.

Molley, J., and J. Edzwald. 1988. Oxidant effects on organic halidate formation and removal of organic halidate precursors. *Environ. Technol. Lett.* 9: 1089–1104.

Parekh, S. 1988. *Reverse osmosis technology application for high purity water production.* New York: Marcel Dekker.

Perrins, J.K., W.J. Cooper, J.H. van Leeuwen et al. 2006. Ozonation of seawater from different locations: Formation and decay of total residual oxidant—Implications for ballast water treatment. *Mar. Pollut. Bull.* 52(9): 1023–33.

Peters, T., and D. Pinto. 2006. Sub-seabed drains provide intake plus pretreatment. *Desalination and Water Reuse* 16, no. 2 (August/September): 23–7.

Rossum, J.R., and D.T. Merrill. 1983. An evaluation of the calcium carbonate saturation indexes. *J. Am. Water Works Assoc.* 75(2): 95–100.

Schock, M.R. 1991. *Internal corrosion and deposition control in water quality and treatment.* New York: McGraw Hill.

Schwartz, J. 2000. *Beach well intakes for small seawater reverse osmosis plants.* Muscat: The Middle East Desalination Research Center.

Seacord, T.F., E. Singley, G. Juby et al. 2003. Post-treatment concepts for seawater and brackish water desalination. In *Proceedings of the Membrane Technology Conference, Atlanta, Georgia.* Denver: American Water Works Association.

Sessions, W.B., and N.S. Hawkins. 2006. Many advantages to working offshore. *Desalination and Water Reuse* 15, no. 4: 22–9.

Voutchkov, N. 2004a. Thorough study is key to large beach well intakes. *Desalination and Water Reuse* 14, no. 1 (May/June): 16–20.

Voutchkov, N. 2004b. Seawater desalination costs cut through power plant co-location. *Filtr. Separat.* 41(7): 24–6.

Voutchkov, N. 2006. Innovative method to evaluate tolerance of marine organisms. *Desalination and Water Reuse* 16, no. 2: 28–34.

Watson, I.C., O.J. Morin, Jr., and L. Henthorne. 2003. *Desalting handbook for planners.* 3rd ed. Denver: U.S. Department of the Interior, Bureau of Reclamation, Technical Service Center, Water Treatment Engineering and Research Group (Desalination and Water Purification Research and Development Program Report No. 72).

WHO. 2008. *Guidelines for drinking-water quality.* 3rd ed. incorporating first and second addenda. Geneva: World Health Organization. http://www.who.int/water_sanitation_health/dwq/gdwq3rev/en/index.html.

Wilf, M., L. Awerbuch, C. Bartels et al. 2007. *The guidebook to membrane desalination technology: Reverse osmosis, nanofiltration and hybrid systems process design, applications and economics.* Rehovot: Balaban Publishers.

3 Chemical Aspects of Desalinated Water

John Fawell, Mahmood Yousif Abdulraheem, Joseph Cotruvo, Fatimah Al-Awadhi, Yasumoto Magara, and Choon Nam Ong

CONTENTS

The internationally recognized compendium of information on the health and safety considerations associated with drinking water is the World Health Organization's (WHO) *Guidelines for Drinking-Water Quality* (GDWQ) (WHO 2008), along with the associated technical documents and guidances produced by WHO. The GDWQ are developed with respect to the more traditional freshwater sources of drinking water and the technologies and approaches for ensuring the healthfulness of the drinking water that is produced. Virtually all of the principles and information in the GDWQ are applicable to desalinated water; however, this assessment was undertaken because there are some additional matters that arise from desalination that are not normally issues for freshwater supplies. This chapter considers the chemicals present in raw source water or introduced during the various stages of producing drinking water from desalinated water.

3.1 CHEMICALS AND DESALINATION

Chemicals occur in the desalination process from numerous origins. These include the source water and chemicals that are used in the treatment process to aid its efficient functioning, to ensure microbiological safety, to stabilize the water before it enters the distribution system, and to control corrosion from contact surfaces during storage and distribution to consumers. Many of these are the same chemicals that would be encountered in conventional drinking water sources and supplies, but there are a number that are of particular relevance for desalination. As the process is based on removal of inorganic salts and also organic chemicals, most of them, including many disinfection by-product (DBP) precursors, will not reach the final water in more than trace quantities and so will not pose a risk to consumers; others will be significantly reduced in concentration. Some low-molecular-weight organic materials will not be entirely removed, and some will be added posttreatment and will therefore reach consumers. Conversely, some of the inorganic ions that are removed may be of nutritional significance and therefore potentially beneficial if they had remained in the finished water.

The different processes used in desalination have some different requirements for chemical additives, and there are also differences in the removal of chemicals from the source water, but there is considerable overlap.

The chemicals of potential interest include

- Natural and anthropogenic chemicals in source water
- Chemicals used or introduced during pretreatment
- Added chemicals and by-products of chemical reactions during drinking water treatment
- Chemicals added to improve the performance of the process
- Chemicals added posttreatment
- Contaminants and corrosion products from contact surfaces

Consideration of any potential risks will depend on a number of factors, not least of which is vulnerable subpopulations that might be exposed. Groups such as dialysis patients are not considered here, because the primary focus is drinking water, and dialysis is a medical intervention that will require further water treatment at the point of use to ensure that it meets very stringent requirements. However, bottle-fed infants are one such group because of their proportionally high intake of water and their developing physiology, which means that they may be less able to handle high concentrations of some natural constituents such as sodium.

In addition, many desalination-dependent locations and countries are in arid regions, where consumption of water may be higher than the norm of 2 L/day for adults, 1 L/day for a 10 kg child, and 0.75 L/day for a 5 kg infant, as used by WHO in developing guidelines for drinking-water quality. Therefore, water consumption rates and dietary habits also need to be considered by health authorities when establishing local standards or guideline values (WHO 2008). Consumption levels of 4 L/day or more would not be unusual for physically active people in warm climates.

As with other water supplies, the approach to ensuring the safety of the use of desalinated water for drinking water should follow the standard water safety plan (WSP) approach applied to an individual supply. This requires identification of the hazards and assessment of the risks, determination that appropriate barriers are in place to remove the hazards or mitigate the risks, and action to ensure that the barriers are always optimized. It also requires that consideration be given as to whether an intervention to control or mitigate risks from one hazard will impact other hazards and risks. In this respect, it is also important to make a comparative assessment of risks so that minor risks are not reduced at the expense of increasing more significant risks (Davison et al. 2005). This is particularly relevant for chemicals in relation to desalination processes.

The following discussion considers some of the chemicals and sources of chemicals that could be present in raw source water or introduced during the various stages of producing desalinated water and its final preparation for drinking water. One potential difference with desalination is the use of seawater or other water sources for blending with the final desalinated water. Water used for blending also needs to be considered and treated appropriately under the WSP, as such water could be a source of contamination of the final drinking water.

3.2 CHEMICALS IN SOURCE WATER

The term *source water* applies to both the saline water that will be the feed to the desalination process and also the potentially saline water source that is used for blending. As with water that will undergo desalination, the level and type of pretreatment applied to blending water will be a significant consideration.

Many inorganic chemicals are found naturally in the seawater or brackish water used as the feed for desalination. These include all of the inorganic chemicals that the process is designed to remove, including sodium chloride, but it will include some that may impact the pretreatment and, potentially, posttreatment stages, such as bromide and, to a lesser extent, borate and iodide.

3.2.1 Boron and Borate

Most of the inorganic components will be significantly removed in the desalination process, either thermal desalination or reverse osmosis (RO) desalination, although some sodium chloride and bromide may be present in the treated water from membrane plants and possibly from some older distillation plants. In terms of key contaminants of direct interest for health and the environment, the most important is probably boron, which can be of significance in RO plants, as the rejection ratio of boron-containing anions (probably mostly as borate) is less than that for most other inorganics.

In the third edition of the GDWQ, the health-based guideline value for boron (borate) in drinking water is 0.5 mg/L (WHO 2008); the Canadian and Australian guidelines are 5 mg/L. In 2009, WHO issued a draft revised guideline value of 2.4 mg/L for public comment (WHO 2009a), and it is expected that WHO will formally issue this guideline value for the fourth edition, expected to be published in 2011 (WHO in preparation). The revised guideline is based on a review of the toxicology data and studies in areas with high background exposures. Although boron is an essential element for plant growth, it is herbicidal at higher levels, and some plants are sensitive at 0.5 mg/L. The latter is the principal issue for residual boron—that is, its effect as an herbicide if present in sufficient amount in irrigation water, particularly in areas where rainfall is so low as to not cause sufficient leaching of salts from soils. Acceptable boron concentrations in desalinated water in areas where desalinated water has significant applications for irrigation may best be determined by authorities on a case-by-case basis reflecting costs, end uses, rainfall, climate, and type of agricultural activity in the area.

3.2.2 Bromide and Bromate

Bromide is initially present in seawater in relatively large amounts (~80 mg/L in some regions), so even high (e.g., >95%) percentage removals will allow some bromide, on the order of 1 mg/L to be present in the finished water. The concentrations of bromide in desalinated water will be approximately proportional to the chloride concentration because of similar removal mechanisms for these analogous anions. Inorganic bromide is also present in many fresh waters, especially groundwaters and coastal aquifers affected by seawater intrusion, at up to milligram per liter levels. FAO/WHO (1988) developed an acceptable daily intake (ADI) for bromide of 1 mg/kg body weight, assuming a 60 kg adult drinking 2 L of water per day with a 20% allocation of the ADI to drinking water could give a health-based reference value in the range of 6 mg/L. A similar conclusion is recommended in the draft WHO background document issued for public comment in 2009 (WHO 2009b). Inorganic bromide in desalinated water would generally not be expected to constitute a threat to health, even if seawater were added in blending; however, a statement from WHO would remove any uncertainty among water producers and users.

If ozonation or other oxidation processes are applied to waters with sufficient residual bromide under appropriate conditions, bromate will be formed at concentrations that will likely exceed the current WHO guideline value of 10 μg/L (WHO 2008). Packaged waters produced by bottling distributed desalinated waters derived from high-bromide source water are often treated by ozonation prior to bottling. This would increase the bromate levels in the bottled water beyond the concentrations in the original distributed water if residual bromide is present. Production of chlorine by electrolysis of seawater will also produce very large amounts of bromate. Bromate is carcinogenic in rats and mice in lifetime tests under high-dose conditions, with cancers in the kidney, thyroid, and testes being observed (WHO 2005a). However, there are indications that small amounts of bromate are metabolized and detoxified following ingestion (Bull and Cotruvo 2006); thus, the current guideline probably overestimates any potential risk at low environmentally relevant exposures. The current WHO bromate guideline did not consider this aspect when the dose–response risk was computed. Studies are under way to generate a physiologically based pharmacokinetic model and a revised risk assessment for ingestion in drinking water.

3.2.3 Sodium and Potassium

Sodium concentrations in seawater are in the range of 10,000–15,000 mg/L, depending on the location. Sodium is an essential nutrient, and there is no health-based WHO guideline value for it, which is normally present in relatively low concentrations in drinking waters derived from freshwater sources, but with significant exceptions. The taste threshold is in the region of 200–250 mg/L, depending on the associated anions. Daily dietary intake may approach 10,000 mg/day for some individuals, which is well above the required daily intake. Sodium is essential for adequate functioning of human physiology, although the requirement of infants for sodium is lower than that for children and adults, and high sodium intake may lead to hypernatremia. This is a problem for bottle-fed infants and is the reason why sodium levels in infant formulas have been reduced significantly over time. There have been concerns expressed about the contribution of sodium intake to increasing hypertension across populations. A number of health officials are concerned about the overall intake of salt from all sources, but particularly food, which is the major source of sodium intake, and are seeking to persuade their populations to decrease salt intake. On the other hand, hyponatremia can be a serious, including fatal, acute risk if significant perspiration causes high loss of sodium, and there is inadequate sodium intake from the total diet. It is probable that the presence of some sodium in drinking water in very warm climates might be beneficial for persons engaging in heavy physical activity.

Seawater, brackish water, and many fresh waters also usually contain potassium. Seawater concentrations are in the region of 450 mg/L. About 98% of the potassium is removed in the desalination process. Potassium is an essential nutrient, and the recommended daily dietary requirement is more than 3,000 mg/day. There is currently no specific WHO guideline value for potassium; residual concentrations in desalinated water will be small and well below any significant contribution to recommended daily dietary intakes.

3.2.4 Magnesium and Calcium

Magnesium and calcium are essential nutrients and are present in seawaters at concentrations of about 1,200–1,700 mg/L and 400–500 mg/L, respectively. They are the principal defining components in "hard water." They are very efficiently removed by desalination, including nanofiltration, but may be added back to finished water by some processes used to stabilize the water and reduce corrosivity. They will be discussed later in this chapter (see Sections 3.5.1 and 3.5.2).

3.2.5 Organic Chemicals Found Naturally in Source Waters

Naturally occurring chemicals include natural organic matter (NOM), such as humic and fulvic acids, and the by-products of algal and seaweed growth, where this growth occurs to a significant extent. Such chemicals can include substances that can have an impact on the odor of the final water, such as geosmin from cyanobacteria, particularly in brackish water, and a range of toxins from a variety of different organisms, including cyanobacteria and dinoflagellates, that can form significant blooms, although these are usually intermittent in nature. The only one of these potential contaminants for which there is currently a WHO guideline value is the cyanotoxin microcystin-LR, which arises from freshwater cyanobacterial blooms and for which there is a provisional guideline value of 1 μg/L (WHO 2008). Desalination processes will significantly control algal toxins. Work is under way to develop drinking water guidelines for other toxins of these types.

The nature of the natural organic molecules is such that most of them have sufficiently high molecular weights and/or low volatilities that they would not be expected to carry over in thermal desalination processes, although the potential for carryover by steam distillation remains a possibility. Volatile organics are usually vented as part of the distillation process. The carryover would be expected to be small, but for substances such as geosmin, which has an odor threshold measured in nanograms per liter; this could still be of concern for the potential acceptability of the final product. Most of the organic molecules are relatively large (e.g., greater than ~200 Da) and would be expected to be excluded by membranes used in desalination; for example, two of the main marine toxins, saxitoxin and domoic acid, have been shown to be rejected by membranes used in desalination (N. Voutchkov, personal communication, 2006). However, low-molecular-weight polar compounds might require further study in that regard. Solvent-type hydrophobic low-molecular-weight neutral organics can pass through membranes to a significant degree.

There is also a significant potential for anthropogenic contamination of source waters, particularly seawater and estuarine waters, as a consequence of discharges from sewage treatment plants and from industry. The contaminants present at a particular site will depend on both the industrial and shipping activities that are present in the wastewater catchment or that discharge directly to sea and the size of the population served. Many of the substances that can reach source waters are covered in the GDWQ and in the associated document, *Chemical Safety of Drinking-Water: Assessing Priorities for Risk Management* (Thompson et al. 2007). A number of potential contaminants reaching drinking water supplies from upstream wastewater discharges, such as pharmaceuticals and hormones, have attracted significant media attention; however, these have been largely shown to not cross desalination membranes (McGuire Environmental Consultants Inc. 2005). The great majority of those molecules would not be expected to be present in the distillate from thermal processes, but there is a potential issue regarding public perception. Providing reassurance of the adequacy of the barriers to the consuming public would be an important step in a WSP. There is also a significant potential for contamination by petroleum hydrocarbons, particularly in regions where there is substantial oil extraction activity. There is the possibility that more volatile substances may be carried over to product water in thermal distillation processes; these include benzene, toluene, xylenes, ethylbenzene (the BTEX compounds), and solvents such as chloroform, carbon tetrachloride, trichloroethene, and tetrachloroethene. These processes are designed to vent those gases during processing, but it is important to confirm that those types of substances are being adequately removed. There may also be potential for those substances, if present in sufficient quantities, to dissolve in RO membranes, migrate through the membranes, and thus appear in finished waters. Although there are health-based drinking water guideline values for all of these substances, the primary issue regarding the BTEX compounds (except for benzene) is the potential for them to cause unacceptable taste and odor at concentrations much lower than the health-based guideline values (WHO 2008). Prevention of source water contamination is the best method to prevent contamination of finished waters. The assessment of potential hazards and risks from pollutants will require an evaluation of the sources and types of pollutant in the local circumstances.

There have also been suggestions of contamination by metals, particularly mercury, in regions of oil production. Data on actual concentrations in feedwaters are very limited; however, there is an existing GDWQ value for inorganic mercury of 6 μg/L (WHO 2008). Mercury also occurs in the form of organomercury compounds, but these substances are hydrophobic, and the main concern relates to accumulation in aquatic organisms rather than in the drinking water.

3.3 PRETREATMENT

Pretreatment of the source water after intake is normally designed to remove contaminants that will interfere with the desalination process, such as by scale formation or membrane fouling. This treatment can include coagulation and filtration or membrane filtration processes that will remove particulate and organic matter, including significantly reducing NOM. A disinfectant/oxidant is normally applied to minimize fouling and to reduce the risk of pathogens carrying over. Chlorine is the usual disinfectant, although other biocides are sometimes used.

Humic and fulvic acids and other related substances that constitute NOM can react with chlorine (and other disinfectants) to produce a wide range of halogenated and oxidized by-products. In the presence of high bromide concentrations, as are found in seawater and many brackish waters, the bromide is oxidized to bromine or hypobromite, which will take part in the halogenation reactions and produce organobromine products as the predominant by-products. Data from studies on the chlorination of seawater show that the yield of DBPs in the form of trihalomethanes (THMs) is dominated by the brominated THMs, particularly bromoform and, to a lesser extent, dibromochloromethane. The WHO guideline values for these two substances are both 100 μg/L, whereas the guideline values for chloroform and bromodichloromethane are 300 and 60 μg/L, respectively, although new data may result in a revision of this last value by WHO. Numerous other organobromine and organochlorine compounds are also formed at low levels, and there are studies under way that continue to identify more of them. There may also be small quantities of iodinated THMs present, but there are no guideline values for these substances; there are limited data on their presence in disinfected fresh waters (Plewa et al. 2004) and some data on their occurrence in disinfected waters with high salt content (Richardson et al. 2003). The levels of other potential chlorination by-products, such as the haloacetic acids, will also be a function of the precursors present. Again, either distillation or membranes will remove most of these DBPs as well as their precursors.

Organonitrogen compounds, particularly dimethylnitrosamine (NDMA) and other nitrosamines (e.g., nitrosodiethyl amine), may be formed during chloramination, especially during distribution, if the appropriate secondary amines are present in the source water or possibly in coagulants. Numerous *N*-chloro amine and amide compounds are undoubtedly formed, but there are limited data on their occurrence and toxicology. There are limited but increasing data on the formation of *N*-nitroso compounds during drinking water distribution, and there appear to be no data on their formation in desalinated seawater. There is also evidence of the formation of nitrosamines in chlorinated wastewater, where there will also be ammonia and secondary amines present. Thus, where chlorinated sewage effluents are likely to impact the raw water, there may be potential for these compounds to volatilize and be carried over into the desalinated product water. NDMA is known to be poorly removed by RO membranes because of its low molecular weight, and it is often treated by advanced oxidation processes in wastewater reuse.

Where hypochlorite is produced by electrolytic generation from seawater/brine with a high bromide level, this will lead to the formation of bromate. Because it is ionic, bromate is not likely to pass through membranes and would not be expected to carry over in thermal systems. Where hypochlorite is allowed to age, there is also a potential for the formation and buildup of chlorate. Chlorate should also be well removed by either distillation or membranes. Its presence in finished water would be due to posttreatment chlorination.

Theoretically, a number of organic contaminants, both raw water contaminants and those resulting from disinfection, could transfer into the product as a consequence of steam distillation. There is a need to determine how important this is and under what circumstances it will take place.

Membranes provide a barrier to most chemical compounds, although not always a complete barrier. The propensity of boron (as borate or boric acid) and also arsenite to pass through membranes raises the question as to what other anions and small neutral organic molecules will pass through membranes. There is a need for more specific data from actual desalination facilities and for specific types of membranes.

Flash distillation desalination plants are sometimes associated with coastal power plants, and an additional potential concern for which there appear to be no firm data is the use of hydrazine in power plants as an oxygen scavenger (see also Chapter 5). Although hydrazine itself is no longer used, alternatives appear to break down to hydrazine. Where these compounds are used, it is important that there be no potential to transfer, through steam leaks, into the desalination stream.

3.4 CHEMICALS FROM TREATMENT PROCESSES

When cleaning agents for membranes are used, they are applied either online or offline, and those chemicals can be present in the system at high concentrations. Therefore, the membranes should be properly flushed before installation and before the system goes back online, and the flushing solutions should be disposed of to waste. Pretreatment of the waste will be necessary, and it is important that this waste stream be disposed of in such a way that it cannot contaminate either source waters or waters that might be used for subsequent blending with desalinated water.

Materials such as piping and contact surfaces in treatment systems and processes that come into contact with drinking water need to be assessed to ensure that no chemicals leach that could cause the WHO guideline values to be exceeded or that no other substances are introduced that could be a hazard to health or adversely impact the acceptability of the final water. Procedures to ensure that this is the case are an important component of the WSP.

3.5 POSTTREATMENT

There are four primary issues concerning the posttreatment water. These relate to blending, remineralization, disinfection, and materials used for storage and transport of the water to the tap. Desalinated water is often blended with other sources that contribute minerals to the final blended water. Seawater as a source for blending has both advantages and disadvantages (see Chapter 4), particularly in terms of corrosion and taste if the blending levels exceed about 1%. In addition, bromide would likely continue to react with residual disinfectants during storage and distribution. However, blending with seawater results in the addition of sodium and some potassium, calcium, magnesium, chlorides, and other salts to drinking water. Consideration should be given to the natural minerals present and whether these will result in finished water not meeting the WHO guidelines (or equivalent national standards) or having unacceptable taste. There is also an issue regarding potential anthropogenic pollutants from a range of sources that need to be considered on a local basis, whenever any external and potentially minimally treated source is used. It is therefore important to take into account potential pollution sources and threats. Disinfection and possibly filtration of the blending water will be necessary if there is any possibility of microbiological contamination (see Chapter 4), in which case similar considerations regarding the formation of by-products in the blending water apply as discussed under pretreatment processes (see Section 3.3). There are currently WHO guidelines for individual THMs, chloroacetic acids, haloacetonitriles, and some other DBPs (WHO 2008). Generally, the NOM content in finished water is very low, and the yield of by-products from final disinfection would be expected to be low as a consequence (McGuire Environmental Consultants Inc. 2004). Chlorine used for disinfection that is generated from brine with high bromide levels may contain significant levels of bromate that could exceed the WHO bromate guideline for drinking water (WHO 2008).

3.5.1 Remineralization

In a number of cases, water is remineralized to reduce its potential for corrosion. Under these circumstances, it is appropriate to consider whether the methods used, such as percolation through limestone, can also increase the concentrations of important nutritional minerals, particularly calcium and magnesium, in the drinking water. Of course, the diet is the principal source of nutrients and minerals, but drinking water may provide supplemental amounts that could be important for some people. WHO expert consultations on calcium and magnesium in drinking water (WHO 2005b, 2006; Cotruvo 2006) concluded that there was evidence of dietary deficiency of both calcium and magnesium in many parts of the world. This would be particularly acute in developing countries and in women, as well as in some sections of the population, such as the elderly, who are also at highest risk of mortality from ischemic heart disease. Hard water and particularly magnesium, a component of hardness, have been negatively (i.e., beneficially) associated with these conditions in a number of epidemiological studies. Although uncertainties about this association remains, in circumstances where a supply is moving from a source that has significant levels of calcium and magnesium to low-mineral desalinated water, it would be appropriate to consider remineralizing with calcium and magnesium salts. Additionally, calcium intake may reduce osteoporosis risk, and magnesium deficiency may also be associated with metabolic syndrome, indicating a prediabetic condition. However, any decision should be taken in conjunction with health and nutrition authorities in the light of total dietary intakes of nutrient minerals. Blending with 1% seawater would provide about 15 mg magnesium/L and about 5 mg calcium/L to the finished water. It is appropriate for WHO and other organizations to continue to consider the importance of calcium and magnesium for protection against ischemic heart disease and to determine the optimum levels of calcium and magnesium and the importance of the calcium-to-magnesium ratio, in order to provide guidance as to the optimum levels of addition, if appropriate. In particular, there are significant considerations with regard to both cost–benefit in particular circumstances and public perception.

Low fluoride intake is also a potential consideration with regard to loss of fluoride from bone and reduced incidence of dental caries. A recommendation of a WHO working group was for a minimum fluoride concentration of 0.2 mg/L, but this recommendation may require examination and confirmatory studies (WHO 2005b). The recommended WHO guideline value for fluoride is 1.5 mg/L, but the optimal value is usually in the range of 0.5–1 mg/L, based on average ambient temperatures and water consumption patterns. The appropriate value provides a balance between the benefits of fluoridation of drinking water and minimizing the occurrence of dental fluorosis. However, use of the guideline value to develop local standards should take into account climate and water consumption, because the guideline value is associated with an intake of 2 L of drinking water per day. This is also a consideration with regard to artificial fluoridation used to protect against dental caries, where this is a significant problem or there is a significant risk that cannot be addressed through other means (WHO 2005b, 2006). Whether to add fluoride to finished water for dental health is a function of the status of tooth decay incidence in the location, diet (sugar consumption levels), and the ready availability and use of dental care in the area throughout the population. These can be determined by appropriate studies in the area.

With regard to sodium levels in the final water, this requires specific consideration of potentially sensitive populations, such as bottle-fed infants.

In addition, other corrosion-inhibiting chemicals, primarily silicates, orthophosphate, or polyphosphate, may be added to the water. Such chemicals are widely used in many parts of the world and are not of direct consequence for health. However, it is important that they be of a suitable quality for addition to drinking water and that there be no contaminants of concern, particularly those covered in the GDWQ, that would make a significant contribution to the concentrations of such contaminants in drinking water. It is also important that they be verified to be always of an appropriate quality. Approval systems for chemicals that specify the quality and acceptable levels of contaminants are available. Guidance on how such systems can and should operate is under consideration by WHO.

3.5.2 Calcium, Magnesium, and Cardiovascular Disease

This issue was examined in detail in three scientific meetings that were generated by this desalination guidance development process. The first was a meeting of experts assembled by WHO in Rome in 2003. The experts' task was to examine the potential health consequences of long-term consumption of water that had been "manufactured" or "modified" to add or delete minerals. Specifically, this was applied to the consumption of desalinated seawater and brackish water, as well as some membrane-treated fresh waters, and their optimal reconstitution from the health perspective. The latter is economically important, because desalinated waters require stabilization by some form of remineralization, often with calcium carbonate (limestone), to control their corrosivity toward pipes and fixtures while in storage and in transit to consumers. That group concluded, among other things, that, on balance, the epidemiological studies indicated that consumption of hard water, and particularly magnesium, is associated with a somewhat lowered risk of certain types of cardiovascular disease (CVD) (WHO 2005b). It also concluded that only a few minerals in natural waters had sufficient concentrations and distribution to support the notion that drinking water might sometimes be a significant supplement to dietary intake. These included calcium, magnesium, selenium, fluoride, copper, and zinc. It recommended that a detailed state-of-the-art review be conducted prior to consideration of the matter in WHO drinking water guidelines.

That report led to the symposium entitled *Health Aspects of Calcium and Magnesium in Drinking Water* (Cotruvo 2006) and a subsequent WHO Expert Meeting (WHO 2006) on the subject. The symposium presented information that large portions of the population are deficient in calcium and magnesium and that water could make important contributions of calcium and magnesium to the daily diet in individuals who had low intakes from other sources. For desalinated water, remineralization methods that include addition of calcium and magnesium are more desirable, because they also contribute nutrient minerals. Seawater blending also adds back magnesium and calcium.

Finally, WHO organized a meeting of experts to further assess drinking water-related epidemiological, clinical, and mechanistic studies that involved calcium or magnesium or hard water that contains calcium and sometimes magnesium (WHO 2006; Cotruvo and Bartram 2009). A large number of studies have investigated the potential health effects of drinking water hardness. Most of these have been ecological studies and have found an inverse (beneficial) relationship between water hardness and cardiovascular mortality. The best correlations were usually with magnesium. Inherent weaknesses in the ecological study design limit the conclusions that can be drawn from these studies. Analytical case–control and cohort studies are more useful than ecological studies for investigating cause-and-effect relationships. Seven case–control studies and two cohort studies of acceptable quality investigating the relationship between calcium or magnesium and CVD or mortality from CVD were identified in the literature. Of the case–control studies, one addressed the association between calcium and acute myocardial infarction and three the association between calcium and death from CVD. None found a positive or inverse correlation between calcium and either morbidity or mortality. Two examined the relationship between magnesium and acute myocardial infarction, finding no association. Five examined the relationship between magnesium and cardiovascular mortality; although some failed to yield statistically significant results, collectively they showed similar trends of reduced cardiovascular mortality as magnesium concentrations in water increased. Statistically significant benefits (where observed) generally occurred at magnesium concentrations of about 10 mg/L and greater. The cohort studies examined the relationship between water hardness (rather than calcium or magnesium content) and CVD or mortality from CVD and found no association (WHO 2006; Cotruvo and Bartram 2009).

The overall conclusion based on identified case–control and cohort studies was that there is no evidence of an association between water hardness or calcium and acute myocardial infarction or deaths from CVD (acute myocardial infarction, stroke, and hypertension). There does not appear to be an association between drinking water magnesium and myocardial infarction. However, the studies do show a negative association (i.e., protective effect) between cardiovascular mortality and drinking water magnesium. Although this association does not necessarily demonstrate causality, it is consistent with the well-known effects of magnesium on cardiovascular function.

3.5.3 Dietary Supplementation

The geographic distribution of the nutrients in source waters used for drinking water production will be varied and inconsistent, so an appropriate diet should be the principal source. In general, drinking water should not be relied upon as a major contributor of significant trace nutrients to daily intake. However, drinking water can provide supplementation to dietary intakes in some locations. Dietary supplementation is widely practiced for general benefit (e.g., vitamin D in milk, vitamin C in drinks, iron and B vitamins and folic acid in bread and other foods). The only beneficial substances added to drinking water in some areas are fluoride with the intent of strengthening dental enamel and reducing the incidence of tooth decay (dental caries) and ferric iron–ethylenediaminetetraacetic acid (EDTA) complex in some dietary iron-deficient areas, and possibly iodine in some areas with high incidence of goiter in Russia.

WHO states that there is clear evidence that long-term exposure to an optimal level of fluoride results in diminishing levels of caries in both child and adult populations and that fluoride is being widely used on a global scale, with much benefit (WHO 2006). However, good dental care, use of fluoride toothpaste, and low sugar consumption are also important dental health factors. Water fluoridation is controversial in some quarters but generally believed by the dental community and many public health officials to be beneficial and without demonstrable risk. Water fluoridation is a matter of national policy. Seawater is naturally low in fluoride, and the fluoride is further depleted by the desalination process. Optimal fluoridation of the desalinated water can be a significant contributor to daily intake and can reduce the incidence of dental caries in some populations, just as it does with fluoridated fresh waters.

3.6 DISTRIBUTION SYSTEMS

Desalinated water is initially more corrosive than many other drinking-water sources, and it is important, as indicated earlier, that the water be stabilized to minimize the corrosion of metallic pipes and fittings used in distribution and in buildings. Where distribution is through tankers, then the potential for corrosion of the tankers must also be considered. The requirement is that corrosion should not give rise to levels of metals that exceed the WHO guidelines (WHO 2008) or add unacceptable appearance or taste or result in physical damage to surfaces in contact with water (McGuire/Malcolm Pirnie 2006). These can include metals from primary distribution and storage, particularly iron, and from plumbing and fittings in buildings, including lead, copper, and sometimes nickel. Iron has, in the past, given rise to problems with discolored water that significantly reduces the acceptability of the water for both drinking and household uses. Cement- or concrete-lined pipes or storage reservoirs can also be damaged from low-pH and low-alkalinity corrosive water. In many cases, a range of coatings and materials will be used to coat pipes or storage reservoirs or for storage tanks in buildings in order to protect against corrosion. It is important that these materials be of a suitable quality for use with potable water; as indicated earlier, approval schemes have an important part to play in ensuring their safety and reducing the potential impact on consumer acceptability. There is a particular consideration in the approval of materials, as in many of these circumstances they will be used at elevated temperatures, which can exacerbate leaching of component metals.

3.7 ADDITIONAL ISSUES

There have been suggestions that drinking very low mineral content (low total dissolved solids) water can lead to a number of adverse effects on humans, particularly on the gastrointestinal tract, even with a diet that provides an adequate level of essential minerals (Kozisek 2005). However, this hypothesis remains controversial in many quarters. In order to resolve this controversy, there is a need to investigate this subject in more detail to determine its significance in a wide range of circumstances, such as those encountered with desalinated and other potentially low-mineral manufactured waters.

Desalination has been used in some parts of the world for many decades, and this experience potentially provides a basis for total diet and water epidemiological studies of various health outcomes, including CVD, osteoporosis, and metabolic syndrome. Such studies, if properly controlled and with proper consideration of potential confounding factors, would be of considerable value in ensuring the safety of desalinated water. WHO will probably recommend that before-and-after studies of acute CVD mortality should be conducted in drinking water supplies that are undergoing changes in calcium and magnesium content (Cotruvo 2009).

Desalinated water may be used for irrigation, and, as indicated earlier, high levels of boron may be toxic to some crops. Suitability for irrigation may also be affected by the low concentration of ions such as calcium and magnesium, which are also important for plant growth (Yermiyahu et al. 2007). Consideration of specific conditions is therefore required if desalinated water is to be used for irrigation, even when this may be on small-scale gardens, which may still be an important source of crops at the village or household level.

3.8 CHEMICAL-RELATED HEALTH RECOMMENDATIONS

- Governments are encouraged to adopt existing systems or establish systems for specifying the appropriateness and quality of additives encountered in desalination or to adopt existing credible recognized standards for those products that would be tailored to desalination conditions.
- Governments are encouraged to adjust their health-based boron drinking water standards based on the latest assessment (e.g., WHO guideline GDWQ, 4th edition). Agricultural applications may lead to different values.
- The guidelines for DBPs should be continually updated as new data are always being developed.
- WHO is encouraged to publish its draft recommendation for bromide for reference purposes as a result of its widespread presence in fresh waters and desalinated waters.
- There is a need to consider sodium and other minerals in water in relation to bottle-fed infants. Guidance is needed on the impact of higher sodium levels for infant formula.
- There should be a reexamination of the low-dose carcinogenic risks of bromate in light of new information under development.
- Countries should carry out dietary and dental care studies, if necessary, to determine the possible and appropriate role of their water supplies, including desalinated water, as a contributor to total dietary intake requirements of fluoride and other minerals.
- There is a need to evaluate the importance of mineral balance in drinking water, particularly calcium and magnesium, with regard to risks of osteoporosis and ischemic heart disease, respectively, as well as fluoride in relation to loss of fluoride from the skeleton.
- Before-and-after epidemiological studies of calcium and magnesium and CVD should be conducted in water supplies that are changing the composition of their drinking water.
- There is a need to examine in more detail the possibility of adverse effects arising from drinking ultra-low-mineral water.
- The impact of desalinated water on agriculture should also be considered in future guidelines for highly processed waters.

DBP formation is an important quality issue in many water projects. Desalinated water has low total organic carbon, so it should not in itself produce significant amounts of DBPs in finished water, depending on the type of postdisinfection and the length and conditions of distribution. However, some practices, such as producing hypochlorite from electrolysis of seawater, will produce significant amounts of bromate and also of organohalogens, and probably particularly brominated organics, to a greater degree than would occur in most conventional freshwater production. Posttreatment blending can be a source of additional DBP precursors.

3.9 CHEMICAL RESEARCH ISSUES

- Determine whether hydrazine is a real problem associated with desalination plants.
- Document especially the occurrence of DBPs in desalination systems from seawater and brackish water. Attention should be given to systems that use blending. Determine the nature and quantities of organobromine DBPs formed in desalinated drinking water.
- Examine the issue of the potential of entrainment of high-molecular-weight compounds in steam in thermal systems and of carryover of volatiles and the performance of methods to reduce carryover. Membrane process performance for low-molecular-weight chemicals should also be quantified.
- Daily drinking water consumption should be examined and quantified in warm climates for use in improved risk–benefit assessments.
- There should be an assessment of the consequences of long-term consumption of desalinated water in areas with long-term experience.
- Consider in more detail the consequences, if any, of long-term consumption of ultra-low-mineral water.
- Safe management and disposal of desalination concentrates should be studied further and guidelines provided.

REFERENCES

Bull, R.J., and J.A. Cotruvo (eds.). 2006. A research strategy to improve risk estimates for bromate in drinking water. *Toxicology* 221 (Special issue, April 17): 2–3, 135–248.

Cotruvo, J.A. 2006. Health aspects of calcium and magnesium in drinking water. *Water Conditioning and Purification* 48, no. 6 (June): 40–4. http://www.wcponline.com/pdf/Cotruvo.pdf.

Cotruvo, J. 2009. Personal communication.

Cotruvo, J., and J. Bartram (eds.). 2009. *Calcium and magnesium in drinking-water: Public health significance*. Geneva: World Health Organization. http://www.who.int/water_sanitation_health/publications/publication_9789241563550/en/index.html.

Davison, A., G. Howard, M. Stevens et al. 2005. *Water safety plans: Managing drinking-water from catchment to consumer*. Geneva: World Health Organization (WHO/SDE/WSH/05.06). http://www.who.int/water_sanitation_health/dwq/wsp0506/en/index.html.

FAO/WHO. 1988. Bromide ion. In *Pesticide residues in food—1988 evaluations*. Geneva, World Health Organization, Joint FAO/WHO Meeting on Pesticide Residues. http://www.inchem.org/documents/jmpr/jmpmono/v88pr03.htm.

Kozisek, F. 2005. Health risks from drinking demineralised water. In *Nutrients in drinking water*, 148–63. Geneva: World Health Organization. http://www.who.int/water_sanitation_health/dwq/nutrientsindw/en/index.html.

McGuire Environmental Consultants Inc. 2004. *Final report—Disinfection byproduct formation in a simulated distribution system: blending desalinated seawater from the Poseidon Resources, Inc., pilot facility with local drinking water sources.* Santa Monica: McGuire/Malcolm Pirnie.

McGuire Environmental Consultants Inc. 2005. *Final report: Pharmaceuticals, personal care products and endocrine disruptors—Implications for Poseidon Resources Corporation's proposed ocean desalination facility in Carlsbad.* Santa Monica: McGuire/Malcolm Pirnie.

McGuire/Malcolm Pirnie. 2006. *Poseidon Resources Corporation corrosion pilot study. Draft final report.* Santa Monica: McGuire/Malcolm Pirnie.

Plewa, M., E. Wagner, S. Richardson et al. 2004. Chemical and biological characterization of newly discovered iodoacid drinking water disinfection byproducts. *Environ. Sci. Technol.* 38(18): 4713–22.

Richardson, S., A. Thurston, Jr., C. Rav-Acha et al. 2003. Tribromopyrrole, brominated acids, and other disinfection by-products produced by disinfection of drinking water rich in bromide. *Environ. Sci. Technol.* 37(17): 3782–93.

Thompson, T., J. Fawell, S. Kunikane et al. 2007. *Chemical safety of drinking-water: Assessing priorities for risk management.* Geneva: World Health Organization. http://whqlibdoc.who.int/publications/2007/9789241546768_eng.pdf.

WHO. 2005a. *Bromate in drinking-water. Background document for development of WHO Guidelines for drinking-water quality.* Geneva: World Health Organization (WHO/SDE/WSH/05.08/78). http://www.who.int/water_sanitation_health/dwq/chemicals/bromide.pdf.

WHO. 2005b. *Nutrients in drinking water.* Geneva: World Health Organization. http://www.who.int/water_sanitation_health/dwq/nutrientsindw/en/index.html.

WHO. 2006. *Meeting of experts on the possible protective effect of hard water against cardiovascular disease, Washington, D.C., USA, April 27–28, 2006.* Geneva: World Health Organization (WHO/SDE/WSH/06.06). http://www.who.int/water_sanitation_health/gdwqrevision/cardiofullreport.pdf.

WHO. 2008. *Guidelines for drinking-water quality.* 3rd ed. incorporating first and second addenda. Geneva: World Health Organization. http://www.who.int/water_sanitation_health/dwq/gdwq3rev/en/index.html.

WHO. 2009a. *Boron in drinking-water. Background document for development of WHO Guidelines for drinking-water quality.* Geneva: World Health Organization (WHO/HSE/WSH/09.04/54). http://www.who.int/water_sanitation_health/dwq/chemicals/Boron_revision3_novemberJKF2008.doc.

WHO. 2009b. *Bromide.* Geneva: World Health Organization. http://www.who.int/water_sanitation_health/dwq/chemicals/bromide.pdf.

Yermiyahu, U., A. Tal, A. Ben-Gal et al. 2007. Rethinking desalinated water quality and agriculture. *Science* 318: 920–1.

4 Sanitary Microbiology of the Production and Distribution of Desalinated Drinking Water

Pierre Payment, Michèle Prévost, Jean-Claude Block, and Sunny Jiang

CONTENTS

4.1 SOURCES AND SURVIVAL OF PATHOGENIC ORGANISMS

The survival of pathogens in the environment has been extensively studied. Whereas some pathogenic organisms are able to persist in source water and finished drinking water for days to months, they generally do not grow or proliferate in water (Fayer et al. 1998; Graczyk et al. 1999; Tamburrini and Pozio 1999; Wait and Sobsey 2001), with a few exceptions (e.g., *Legionella*, *Vibrio*, *Naegleria*, and *Acanthamoeba*). The survival of these pathogens is strongly related to water temperature, solar radiation, osmotic pressure, and the abundance and activities of predators (e.g., protozoa). Increased water temperature results in an active development of the indigenous flora and fauna, which utilize many microorganisms, including human pathogens, as food sources; this results in the accelerated removal of pathogens in warmer waters compared with waters below 15°C. Sunlight (ultraviolet light) has been identified as an important factor in the inactivation of pathogens (Sinton et al. 1999; Fujioka and Yoneyama 2002). In addition, most human enteric bacteria also decay more rapidly in saline water than in fresh water as a result of the higher osmotic pressure (Nasser et al. 2003). Coastal currents and surface waves can transport pathogens over long distances (Kim et al. 2004; Reeves et al. 2004).

In addition to pathogens associated with anthropogenic sources, some indigenous marine bacteria, such as those belonging to the *Vibrio* genus *(V. cholerae, V. varahaemolyticus, V. vulnificus, V. minicus*), and toxin-producing algae are potential sources of waterborne human health risk. Harmful algal blooms are sources of paralytic, neurotoxic, and amnesic shellfish poisoning. Toxin-producing algal species include the red tide organism *Karenia brevis* (previously named *Gymnodinium breve,* producer of neurotoxic shellfish poisoning), *Alexandrium* sp. (producer of paralytic shellfish poisoning), Pseudo-nitzschia (producer of amnesic shellfish poisoning), *Pfiesteria piscicida*, brown tide organisms (*Aureococcus* and *Aureoumbra*), and cyanobacteria (blue-green algae). *Vibrio cholerae* can be found in waters contaminated by human excreta from populations where the infection is present and constitute a significant health risk in populations using these waters. Indigenous bacterial pathogens and toxin-producing algae display strong seasonal and temporal variability generally due to rapid multiplication when water is warming during spring and summer months and with the availability of additional nutrients. Temperature, salinity, and nutrient sources will have effects on the occurrence of vibrios and algal blooms (Jiang et al. 2000; Jiang and Fu 2001; Louis et al. 2003). Blooms of toxin-producing algae can be triggered by coastal upwelling that brings in nutrient-rich deep water or by anthropogenic nutrients from waste discharge to coastal waters (Li and Daler 2004; Yang and Hodgkiss 2004).

Groundwaters are usually less subject to pathogen contamination owing to the effect of soil filtration. However, the efficiency of pathogen removal during infiltration is largely due to the characteristics of soil substrates. Paul and colleagues (Paul et al. 1995a, 1995b, 1997, 2000) demonstrated that in the Florida Keys, where porous limestone is the major constituent of soil substrate, microbial tracers injected into the groundwater injection wells that were used for standard sewage disposal migrate rapidly through ground aquifers and seep through the bottom of the seafloor to surrounding seawater. Human enteric viruses are found frequently in the canal and marine waters around the Florida Keys as a result of seepage from septic tanks and groundwater injection wells (Griffini et al. 1999; Lipp et al. 2002).

4.2 MONITORING FOR PATHOGENS AND INDICATOR ORGANISMS

Pathogen monitoring has been the subject of numerous discussions. The current consensus is that routine or occasional sampling of source waters for pathogens is not a very efficient approach for managing drinking water quality and treatment processes. The World Health Organization's (WHO) *Guidelines for Drinking-Water Quality* (GDWQ) (WHO 2008) provide guidance and a framework for safe drinking water through water safety plans (WSPs) that include monitoring for various parameters. In that framework, gathering knowledge and data on pathogens in the environment is an integral part of the process. As more emerging pathogens are identified (Cotruvo et al. 2004; AWWA 2006), data obtained by researchers provide useful information on various parameters that can be used as index, indicator, or surrogates for waterborne pathogens. Routine monitoring of source water may be needed to understand the quality of the source water, for consideration in the desalination treatment process utilizing that source water and for an integrated management approach to the watershed. Fortunately, desalination processes should be particularly effective at removing undesirable microorganisms of all types prior to their reaching finished drinking water. Indicator monitoring and pretreatment would be a particular concern for blending waters that may not receive treatment. In addition, pathogen monitoring of watersheds is useful for research-driven ecological modeling initiatives. Routine pathogen monitoring is generally not justified, and a good combination of microbial, indicator, and nonmicrobial parameters is probably a more cost-effective monitoring solution (Dufour et al. 2003).

4.3 MICROBIAL CONSIDERATIONS FOR DESALINATION PROCESSES

The efficiency of desalination plants in removing or inactivating microbial contaminants can be assessed by examining the expected performance and factors affecting the quality of each flow that may be produced or combined into the final treated water. The potential for survival of microorganisms depends on the capabilities and operating conditions of the process units for their removal and/or inactivation. The issue of disinfection or elimination of microorganisms of health concern can be evaluated for the water from the pretreatment stage, for the water produced by membrane processes or the water resulting from thermal treatment processes, and especially for posttreatment blending waters.

4.3.1 PRETREATMENT

The specifics of pretreatment are usually determined by the type of downflow process used for the removal of dissolved solids. For example, membranes need to be protected from particulates to prevent clogging and fouling. The objective of pretreatment for microbials using oxidants and biocides is to prevent fouling of the reverse osmosis (RO) membranes and does not specifically address disinfection goals, because oxidant doses applied are not sufficient to reach residual concentrations required for efficient disinfection.

Other pretreatments include the use of membranes to prepare the water for the subsequent desalination process (see Chapter 2). These pretreatment membranes include microfiltration and nanofiltration (NF) membranes, and they have a substantial capacity to physically remove a considerable proportion of particulate-associated microorganisms (see Chapters 1 and 2) as well as some dissolved solids. They can effectively remove at least 6 logs of microorganisms according to their pore size distribution, but the actual removal should be validated before application as a pretreatment (LeChevallier and Kwok-Keung 2004). If partial flow of these pretreated waters is used to blend with the final processed desalinated water, then an evaluation of the disinfection/removal requirements for this added flow must be completed in order to attain the same level of protection as the one attained by the treatment itself. Dilution alone is unacceptable.

4.3.2 Blending Source Water with Desalinated Water

The quality of the blended water is particularly relevant if there will be mixing of incompletely treated water with desalinated water prior to distribution. This is of primary importance to the evaluation of the microbial risk of the blended water as well as to the consequences of the formation of disinfection by-products (DBPs). The amount of blended water may vary from less than 1% to 10% and can include partially treated seawater and untreated groundwater. Short-circuiting of the treatment process should not allow pathogens and other undesirable microorganisms to be mixed with the finished desalinated water. General guidelines for drinking water and guidelines specific to the minimization of microbial risks should be applied in the context of a complete system assessment to determine whether the drinking water supply (from source through treatment to the point of consumption) can deliver water that meets the health-based targets (LeChevallier and Kwok-Keung 2004; WHO 2008). In addition, there should be specific regulations in each country specifying the minimum requirements for disinfection and particle removal and monitoring methods for appropriate performance surrogates. The required performance for the removal of bacteria, viruses, and protozoan parasites should also be adjusted to the level of contamination of the raw water used for blending. In addition, if the blended water contains high total organic carbon (TOC) or bromide, then chemical disinfection by chlorine or ozone might not be advisable, owing to formation of undesirable by-products.

WHO and other organizations provide guidance and information on removals of bacteria, viruses, and protozoa that are achieved by typical and enhanced water treatment processes. Table 4.1 provides an example of the performance of several common disinfectants to achieve various logs of inactivation of viruses in drinking water using concentration × time (CT) values as the measure and viruses as the challenge (USEPA 1999). These CT values are based on water temperatures of 10°C and pH in the range of 6–9. CT values for chlorine are based on a free chlorine residual, and chlorine is less effective as the pH increases from 6 to 9. At a given CT value, a lower concentration and longer contact time are more effective than the converse. For all of these disinfectants, the effectiveness increases as the temperature increases. CTs would be lower for vegetative bacteria than for viruses, whereas CTs for protozoa (more resistant to disinfection) would be higher. Spores would be much more resistant than vegetative bacteria. Additional performance information is available for various treatment methods and conditions in USEPA (1991, 1999) and the WHO GDWQ (WHO 2008).

TABLE 4.1
CT Values for Inactivation of Viruses

	Inactivation (mg·min/L)		
Disinfectant	**2 log**	**3 log**	**4 log**
Chlorine[a]	3	4	6
Chloramine[b]	643	1067	1491
Chlorine dioxide[c]	4.2	12.8	25.1

[a] Based on temperature 10°C, pH 6–9, free chlorine residual of 0.2–0.5 mg/L.
[b] Based on temperature 10°C, pH 8.
[c] Based on temperature 10°C, pH 6–9.
Source: CT values from USEPA 1991. *Guidance manual for compliance with the filtration and disinfection requirements for public water systems using surface water sources*. Washington: U.S. Environmental Protection Agency (EPA No. 570391001). http://www.epa.gov/safewater/mdbp/guidsws.pdf.

TABLE 4.2
CT Values for Inactivation of Viruses Using Chloramines

	Inactivation (mg·min/L)		
Temperature (°C)	**2 log**	**3 log**	**4 log**
5	857	1423	1988
10	643	1067	1491
15	428	712	994
20	321	534	746
25	214	356	497

Source: CT values from USEPA 1991. *Guidance manual for compliance with the filtration and disinfection requirements for public water systems using surface water sources*. Washington: U.S. Environmental Protection Agency (EPA No. 570391001). http://www.epa.gov/safewater/mdbp/guidsws.pdf.

Table 4.2 provides an illustration of the temperature dependence of CT values for disinfection using viruses and chloramines as the example. Inactivation improves significantly with increasing temperature. Even so, chloramines are much less effective than chlorine or chlorine dioxide, but they have the benefit of being more persistent during storage and distribution.

In addition, there could be specific regulations in each country specifying the minimum requirements for disinfection and particle removal. If the blended water contains high TOC or bromide, the potential for the formation of additional DBPs, such as bromate and organochlorinated and organobrominated by-products, should be evaluated when selecting the best processes to achieve the performance target.

4.4 REVERSE OSMOSIS

RO has been shown to remove bacteria and larger pathogens and all or a large fraction of viruses. Even if the primary application of RO membranes is desalination, RO and NF are increasingly being applied to treat surface water with the objective of removing pesticides, precursors of DBPs, and pathogens (Gagliardo et al. 1997; Adham et al. 1998a; van der Hoek et al. 2000). With the performance reported at bench and full scale, high-quality RO processes are good treatment barriers to pathogens if properly selected and maintained.

WHO provides guidance on target removals for bacteria, viruses, and protozoa, removals that are achieved by typical and enhanced water treatment processes (WHO 2008). Removal of viruses by RO membranes may vary significantly and is a function of the membrane itself as well as its condition and the integrity of the entire system, including seals. Removals ranging from 2.7 to more than 6.8 logs, depending on the type of RO membrane, have been reported at bench scale using MS2 bacteriophage as the model virus, and the authors suggested that the selection of membranes is an important factor in determining virus removal (Adham et al. 1998b). Kitis et al. (2002, 2003) reported removals of MS2 ranging from 5 logs for a dual-element unit to more than 6.8 logs for a multistage unit. Pilot-scale studies were conducted to investigate the potential of integrated ultrafiltration (UF) and NF membrane systems for the removal of various microorganisms, including viruses, protozoa (*Cryptosporidium* oocysts and *Giardia* cysts), bacterial spores (*Clostridium perfringens*), and bacteriophage (MS2 and PRD-1). Lovins et al. (1999) observed removals, including those resulting from pretreatment, ranging from 6.1 to 10.1 logs, showing that membrane treatment exceeds the microbial removal attained by other combinations of process units, such as coagulation, filtration, and disinfection of surface water. As microorganisms are frequently bound to particulates, some turbidity might actually improve the removals by membranes with larger nominal pore sizes.

4.4.1 Integrity of the RO System

Although RO constitutes an excellent barrier to microorganisms, the maintenance of that barrier depends on the integrity of the system. Breaches of integrity in the membranes or the O-rings could lead to the passage of pathogens into the process water and must be monitored by integrity testing. Building on bench-scale studies done by Colvin et al. (2000), Kitis et al. (2002) critically compared three integrity-testing methodologies at pilot scale. They investigated the ability of these tests (1) to quantify virus removal (MS2 bacteriophage) in single-element and two-stage configurations and (2) to determine the changes in virus removal capability when systems are subject to different types of membrane and gasket compromising and fouling. These authors concluded that the loss of membrane integrity decreased virus removal from 5.3 to 2.3 logs when the compromised unit was placed in the lead position and from 5.3 to 4.2 logs when the compromised unit was in the trailing position. Fouling appeared to limit the impact of imperfections by a combination of cake formation and pinhole filling. Cracking of the O-rings did not lead to significant decreases in the removals of MS2 or indicators, and the location of the damage influenced the extent of the small decrease in performance. They also concluded that Rhodamine WT could be used as an indicator of virus removal, but its feasibility at full scale was questioned.

Effective methods to measure the integrity of RO membranes should be used to achieve target removals (WHO 2008). Currently, conductivity measurements are utilized, but the sensitivity limits their application to about 2 logs of removal. Online measurement of integrity using safe, easily detected, high-molecular-weight chemicals such as Rhodamine WT and new developments in biomonitoring may eventually provide innovative tools to ensure adequate performance monitoring.

Bacteria have been found in permeate samples of NF and RO effluent, and they can proliferate in discharge lines. This does not mean that pathogens are not rejected, but rather that sterile conditions cannot be maintained (Taylor and Jacobs 1996). As bacteria have been shown to traverse through membrane defects, membranes cannot be considered as completely effective for disinfection and are commonly succeeded by a disinfection step.

4.4.2 Fouling and Biofouling

As bacteria can migrate through membrane defects, passage as well as microbial growth (biofouling) may affect the quality of the product. Of potential concern is the impact of biofilm growth on membranes and the potential for retaining pathogens and growth of bacterial pathogens. Biofilms can negatively impact membrane integrity directly or positively by plugging holes and imperfections.

Fouling of membranes is the progressive accumulation of material on the membrane surface or in its pores. The influence of the quality of feedwater on fouling, notably the characteristics and composition of natural organic matter, the pretreatment applied, the types of coagulants, and the membrane characteristics, has been studied intensively (Lahoussine-Turcaud et al. 1992; Ridgway and Flemming 1996; Carroll et al. 2000; Cho et al. 2000; Fan et al. 2001; Fonseca et al. 2003; Gabelich et al. 2003; see Chapter 2 for additional information).

Biofouling occurs when microorganisms accumulate and/or grow on the membrane surface, resulting in a premature decrease of flux through the membrane and/or increase in pressure drops. According to Flemming et al. (1997), practically every membrane system operating with water supports biofilms, but not all systems experience operational problems because of excessive biofilm formation. Fifty-eight of 70 RO plants in the United States reported biofouling problems (Paul 1991). In a survey of NF and RO treatment plants in the Netherlands, Vrouwenvelder and van der Kooij (2001) observed biofouling problems in 12 of the 13 plants studied.

There are several reasons why it is important for drinking water suppliers to avoid growth in distribution systems, including regulatory requirements, aesthetic and organoleptic acceptability, and the difficulty in maintaining a residual disinfectant (Prévost et al. 2005). Regulatory issues associated with bacterial growth include interference between the high abundance of background heterotrophic bacteria and coliform detection when lactose-based growth media are used, violation of heterotrophic plate count (HPC) regulations in some countries, and violation of coliform rules. The development of HPCs during distribution is no longer considered a significant health risk per se, but its value as an indicator of water quality and treatment has been reiterated (WHO 2003).

4.5 ORGANIC MATTER AND GROWTH OF MICROORGANISMS IN DESALINATED WATER

Desalinated water typically has low or very low organic carbon concentrations (TOC) and also most likely very low biodegradable organic matter (BOM) concentrations. NF and RO appear to be the most efficient processes available for the removal of BOM at this time. Laurent et al. (2005) provided a summary of dissolved organic carbon and BOM removals by RO. Even if RO has excellent potential for significant removal of BOM, the documented removals by full-scale RO membranes are rare (Escobar et al. 2000). Hong et al. (2005) reported that assimilable organic carbon (AOC) removals by a hybrid NF–RO system demonstrate that low levels of low-molecular-weight AOC pass through. Although not reporting actual levels of BOM, MacAree et al. (2005) reported bulk water densities in HPCs (<6 colony-forming units/mL) and total direct counts by epifluorescence microscopy (<6,000 cells/mL) typical of starved systems.

The impact of BOM on the potential for the establishment and subsequent development of opportunistic and frank pathogens in biofilms fixed on RO membranes has not been quantified. However, some information is available on the colonization and survival of pathogens on filter media, activated carbon fines, pipe surfaces, point of entry, point of use, softening systems, cooling towers, and air conditioning systems. Overall, indications are that nonparasitic pathogens are not competitive in establishing and developing in mixed heterotrophic indigenous biofilms, with the noted exception of *Legionella*, as the presence of microbial contaminants in RO-treated water would be mostly related to breakthrough rather than to multiplication/colonization in the system itself (Laurent et al. 2005). Furthermore, occurrence in the distribution system could also be due to ingress of contaminated water (Ainsworth 2004).

The passage of bacterial pathogens from the biofilm or the pretreated water depends on the availability of orifices in the membranes and O-rings. These issues do not differ from the general issues of membrane integrity that must be addressed to prevent the passage of pathogens in the bulk liquid. Some bacteria have been shown to survive the RO membrane process and to be carried throughout the distribution system. Identified species include capsulated bacteria such as *Novosphingobium capsulatum* (MacAree et al. 2005).

4.6 THERMAL PROCESSES

When thermal processes are used for desalination, microbial inactivation will be controlled by the temperature attained and the time the water remains at that temperature. Typical temperatures to ensure the inactivation of vegetative cells by humid heat vary from 50°C–60°C when maintained for 5–30 min to achieve pasteurization. Spores, endospores, and other resistant forms are more resistant to heat and require higher temperatures (70°C–100°C) held for longer periods of time. Most vegetative pathogens are inactivated under flash pasteurization conditions (temperature of 72°C for 15 s). The condensate is unlikely to contain pathogens after the distillation process because of the killing impact of heat and because pathogens are unlikely to be entrained. However, reduced pressures are used in some desalination processes to reduce the boiling point and reduce energy demand. Temperatures as low as 50°C may be utilized (USBR 2003) and might not achieve the required inactivation targets. Inactivation levels expected at temperatures typical of distillation processes are considered sufficient to inactivate most pathogens, as they are equivalent to or in excess of those used for pasteurization.

4.7 DISINFECTION OF DESALINATED WATERS

Desalinated waters constitute a relatively easy disinfection challenge because of their low TOC and particle content, low microbial loads, and minimal oxidant demand after desalination treatments. Turbidity is not likely to affect chemical disinfectant performance, as turbidity values of desalinated water are relatively low. Posttreatment (e.g., lime) can cause an increase of inorganic turbidity that would not interfere with disinfection. The target levels of inactivation for pathogens remaining in desalinated waters can readily be achieved by appropriate disinfection processes, discussed elsewhere (WHO 2008). Once the target levels of disinfection have been achieved and as part of a WSP, an appropriate level of chlorine-based residual disinfectant should be maintained during distribution.

Issues to be considered as specific to the disinfection of desalinated water are

- The potential passage of viruses through some RO membranes, which brings the point of adequate virus inactivation requirements downstream of RO.
- The potential loss of integrity of membranes, which could lead to the passage of various pathogens in process water.
- The practice of blending nondesalinated water to remineralize the treated water, which raises the need to define appropriate targets for treatment and disinfection of the water used for blending.

Except for blending water that should usually be treated, these issues can be addressed in most cases by applying effective postdesalination disinfection using chlorine-based or alternative disinfection processes (ultraviolet light, ozone, etc.) as an additional barrier in a risk reduction WSP approach (WHO 2008).

4.8 STORAGE AND DISTRIBUTION OF PROCESSED WATER

The challenge of maintaining water quality during storage and distribution is not specific to desalinated water. Microorganisms will grow during distribution, especially in the absence of an effective residual disinfectant and at the high water temperatures often encountered (Block 1992). A broad spectrum of microbial species, such as *Legionella*, *Aeromonas*, *Pseudomonas*, *Burkholderia pseudomallei,* and atypical mycobacteria, some of which include opportunistic pathogen strains, can be present in distributed waters. The routes of transmission of these bacteria include inhalation and contact (bathing), with infections occurring in the respiratory tract, in skin lesions or in the brain (Craun and Calderon 2001). There is no evidence of an association of any of these organisms with gastrointestinal infection through ingestion of drinking water (Ainsworth 2004), but *Legionella* can grow to significant numbers at temperatures of 25°C–50°C. Water temperature is an important element of the control strategies. Wherever possible, water temperatures should be kept outside the range of 25°C–50°C. In hot water systems, storage should be maintained above 55°C, and similar temperatures throughout associated pipework will prevent growth of the organism. However, maintaining temperatures of hot water above 50°C may represent a scalding risk (WHO/WPC 2006). Where temperatures in hot or cold water distribution systems cannot be maintained outside the range of 25°C–50°C, greater attention to disinfection and strategies aimed at limiting development of biofilms are required. Accumulation of sludge, scale, rust, algae, or slime deposits in water distribution systems supports the growth of *Legionella* spp., as does stagnant water (Lin et al. 1998). Systems that are kept clean and flowing are less likely to support excess growth of *Legionella* spp. Care should also be taken to select plumbing materials that do not support microbial growth and the development of biofilms. The WHO (2008) guidelines and WHO/WPC (2006) provide more information on the subject.

The maintenance of water quality during storage and distribution depends on a number of factors, including the following:

- The amount of biodegradable organic matter available and trace nutrients to support the growth of suspended and fixed bacteria
- The chemical balance to limit the release of iron, lead, and copper
- The maintenance of an oxidant residual
- The availability and nature of attachment surfaces, in particular, the pipe and reservoir surfaces, and the presence of corrosion
- The maintenance of integrity in the pipes and reservoirs
- The growth conditions, such as residence time, hydraulic conditions, and temperature

The WHO documents *Safe Piped Water* (Ainsworth 2004) and *Health Aspects of Plumbing* (WHO/WPC 2006) set risk management and reduction frameworks to limit the health risk associated with the distribution of piped water, and these guidelines also apply to desalinated water. Those water quality concerns should be considered in light of the potential for microbial regrowth, DBP formation, and the control of pipe corrosion.

High water temperatures will limit the maintenance of an effective residual throughout the distribution system as a result of the increased chemical reactivity of the disinfectant. The use of chloramines constitutes an advantageous alternative to free chlorine in distribution systems with long residence times and elevated temperatures. Chloramines also seem to be more effective at limiting *Legionella* growth in domestic plumbing; however, nitrification can occur from chloramines when *Nitrosomonas* bacteria are present.

4.9 ISSUES WITH BLENDING PRODUCT WATER WITH OTHER SOURCES

Blending of the desalinated water with groundwater or other sources of potable water may be utilized. This practice does not raise any special issues for desalinated water, with the exception that special care should be taken regarding the chemical stability of water to prevent the release of iron, lead, and copper. Taylor et al. (2006) provided an excellent review of issues to take into consideration and results from an extensive pilot study of the impact of blending RO-treated water with potable water from groundwater and surface water in various pipe materials. Of special relevance are the maintenance of conditions to minimize iron, lead, and copper release (selection of disinfectant and dosage adjustment) and the control of nitrification when chloramines are used as the secondary disinfectant.

4.10 RECOMMENDATIONS

As with all drinking water supplies, desalinated drinking water production should strive to utilize the best available source water. The desalination plants should be located away from sewage outfalls, storm drains, and areas with recurring harmful algal blooms. The level of impact from sewage discharge will depend on the local conditions, but special attention should be given to the location of water intakes and wastewater outfalls, especially if costing is considered. Suitable monitoring of biological and physiochemical parameters of source water during plant operation will ensure that the treatment processes are not overcome by high levels of pollutants.

Monitoring of source water quality for pathogens in a desalination process is not an effective operational approach owing to issues similar to those encountered in monitoring of freshwater sources. However, it is useful to provide baseline information on microbial quality that is capable of indicating significant changes in water quality in order to ensure treatment efficiency. This baseline information includes various biological and physicochemical parameters that are described in Table 5.1 in Chapter 5. Information on potential risks to human health can be obtained through use of indicators of fecal pollution such as *Escherichia coli* and enterococci, which are suggestive of the presence of pathogens. Furthermore, as enterococci are more resistant than *E. coli* to environmental degradation conditions in the marine environment (WHO 2003), they are likely to be a better indicator of fecal pollution in seawater and the presence of enteric pathogens. Coliphages, which are viruses that infect coliform bacteria and are similar to human viruses in respect to survival and decay in seawater, have also been suggested as useful indicators of source water quality as well as treatment process performance. HPCs are not an independent indicator of fecal pollution, yet they may be used as a baseline parameter to indicate treatment effectiveness and microbial community changes. Development and proper applications of analytical methods better adapted to saline water are needed to improve the monitoring of saline source waters. Some specific recommendations include the following:

- Maintain final disinfection after desalination to ensure the inactivation of bacteria and viruses and maintenance of a residual during storage and distribution.
- Water used for blending should be treated to reach microbial quality goals set on the basis of raw water contamination and risk reduction. Contaminated water should not be blended with desalinated process water.
- Treatment should be designed to ensure the presence of multiple barriers, including a final disinfection barrier.
- Maintenance of water quality during storage and distribution, including the presence of a disinfectant residual, is important to ensure that product water quality is maintained to the consumer.
- WHO guidelines applicable to storage and distribution systems should be applied to minimize growth and recontamination.
- Desalinated water is very low in nutrients and possesses a low microbial growth potential. However, high temperatures (30°C–45°C) that are frequent in some countries using desalination may enhance the growth of pathogens such as *Legionella,* and nitrification processes when chloramines are used.

REFERENCES

Adham, S., P. Gagliardo, D. Smith et al. 1998a. Monitoring the integrity of reverse osmosis membranes. *Desalination* 119(1–3): 143–150.

Adham, S., R.S. Trussell, P.F. Gagliardo et al. 1998b. Rejection of MS-2 virus by RO membranes. *J. Am. Water Works Assoc.* 90(9): 130–5.

Ainsworth, R. (ed.). 2004. *Safe piped water. Managing microbial water quality in piped distribution systems.* Geneva: World Health Organization. http://www.who.int/water_sanitation_health/dwq/924156251X/en/index.html.

AWWA. 2006. *Waterborne pathogens.* Denver: American Water Works Association (AWWA Manual of Practices, M48).

Block, J.C. 1992. Biofilms in drinking water distribution system. In *Biofilms—science and technology*, ed. L.F. Melo, T.R. Bott, M. Fletcher et al., 469–85. Dordrecht: Kluwer Academic Publishers.

Carroll, T., S. King, S.R. Gray et al. 2000. The fouling of microfiltration membranes by NOM after coagulation treatment. *Water Res.* 34(11): 2861–8.

Cho, J., G. Amy and J. Pellegrino. 2000. Membrane filtration of natural organic matter: Factors and mechanisms affecting rejection and flux decline with charged ultrafiltration (UF) membrane. *J. Memb. Sci.* 164: 89–110.

Colvin, C.K., C.L. Acker, B.J. Marinas et al. 2000. Microbial removal by NF/RO. Presented at the American Water Works Association Annual Conference, Denver, Colorado, June 11–15, 2000.

Cotruvo, J.A., A. Dufour, G. Rees et al. (eds.). 2004. *Waterborne zoonoses: Identification, causes and control.* London: IWA Publishing on behalf of the World Health Organization (Emerging Issues in Water and Infectious Disease Series). http://www.who.int/water_sanitation_health/diseases/zoonoses/en/index.html.

Craun, G.F., and R.L. Calderon. 2001. Waterborne disease outbreaks caused by distribution system deficiencies. *J. Am. Water Works Assoc.* 93(9): 64–75.

Dufour, A., M. Snozzi, W. Köster et al. 2003. *Assessing microbial safety of drinking water: Improving approaches and methods.* London: IWA Publishing on behalf of the World Health Organization (Drinking Water Quality Series). http://www.who.int/water_sanitation_health/dwq/9241546301/en/.

Escobar, I.C., S. Hong and A.A. Randall. 2000. Removal of assimilable organic carbon and biodegradable dissolved organic carbon by reverse osmosis and nanofiltration membranes. *J. Memb. Sci.* 175(1): 1–18.

Fan, L., J.L. Harris, F.A. Roddick et al. 2001. Influence of the characteristics of natural organic matter on the fouling of microfiltration membranes. *Water Res.* 35(18): 4455–63.

Fayer, R., T.K. Graczyk, E.J. Lewis et al. 1998. Survival of infectious *Cryptosporidium parvum* oocysts in seawater and eastern oysters (*Crassostrea virginica*) in the Chesapeake Bay. *Appl. Environ. Microbiol.* 64(3): 1070–4.

Flemming, H.C., G. Schaule, T. Griebe et al. 1997. Biofouling—the Achilles heel of membrane processes. *Desalination* 113(2): 215–25.

Fonseca, A.C., S.C. Summers, A.R. Greenberg et al. 2003. Isolating and modeling critical factors governing biofouling of nanofiltration membranes. Presented at the American Water Works Association Membrane Technology Conference, Atlanta, Georgia.

Fujioka, R.S., and B.S. Yoneyama. 2002. Sunlight inactivation of human enteric viruses and fecal bacteria. *Water Sci. Technol.* 46(11–12): 291–5.

Gabelich, C.J., T.I. Yun, B.M. Coffey et al. 2003. Pilot-scale testing of reverse osmosis using conventional treatment and microfiltration. *Desalination* 154: 207–23.

Gagliardo, P.F., S. Adham, R.R. Trussell et al. 1997. Membranes as an alternative to disinfection. Presented at the American Water Works Association Annual Conference, Atlanta, Georgia.

Graczyk, T.K., R.C. Thompson, R. Fayer et al. 1999. *Giardia duodenalis* cysts of genotype A recovered from clams in the Chesapeake Bay subestuary, Rhode River. *Am. J. Trop. Med. Hyg.* 61(4): 526–9.

Griffini, O., M.L. Bao, K. Barbieri et al. 1999. Formation and removal of biodegradable ozonation by-products during ozonation–biofiltration treatment: pilot-scale evaluation. *Ozone Sci. Eng.* 21(1): 79–98.

Hong, S., I.C. Escobar, J. Hershey-Pyle et al. 2005. Biostability characterization in a full-scale hybrid NF/RO treatment system. *J. Am. Water Works Assoc.* 97(5): 101–10.

Jiang, S.C., and W. Fu. 2001. Seasonal abundance and distribution of *Vibrio cholerae* in coastal waters quantified by a 16S–23S intergenic spacer probe. *Microb. Ecol.* 42(4): 540–8.

Jiang, S.C., V. Louis, N. Choopun et al. 2000. Genetic diversity of *Vibrio cholerae* in Chesapeake Bay determined by amplified fragment length polymorphism fingerprinting. *Appl. Environ. Microbiol.* 66(1): 140–7.

Kim, J.H., S.B. Grant, C.D. McGee et al. 2004. Locating sources of surf zone pollution: A mass budget analysis of fecal indicator bacteria at Huntington Beach, California. *Environ. Sci. Technol.* 38(9): 2626–36.

Kitis, M., J.C. Lozier, J.H. Kim et al. 2002. Microbial removal and integrity monitoring of high-pressure membranes. Presented at the American Water Works Association Water Quality Technology Conference, Seattle, Washington.

Kitis, M., J.C. Lozier, J.H. Kim et al. 2003. Evaluation of biologic and non-biologic methods for assessing virus removal by and integrity of high pressure membrane systems. *Water Supply* 3(5–6): 81–92.

Lahoussine-Turcaud, V., M. Wiesner, J.Y. Bottero et al. 1992. Coagulation–flocculation with aluminium salts: Influence on the filtration efficacy with microporous membrane. *Water Res.* 26(5): 695–702.

Laurent, P., P. Servais, V. Gauthier et al. 2005. Biodegradable organic matter and bacteria in drinking water distribution systems. In *Biodegradable organic matter in drinking water treatment and distribution*, ed. M. Prévost, P. Laurent, P. Servais et al., 147–204. Denver: American Water Works Association.

LeChevallier, M.W., and A. Kwok-Keung. 2004. *Water treatment and pathogen control: Process efficiency in achieving safe drinking water*. London: IWA Publishing on behalf of the World Health Organization. http://www.who.int/water_sanitation_health/dwq/9241562552/en/index.html.

Li, D., and D. Daler. 2004. Ocean pollution from land-based sources: East China Sea, China. *Ambio* 33(1–2): 107–13.

Lin, Y.E., R.D. Vidic, J.E. Stout et al. 1998. *Legionella* in water distribution system. *J. Am. Water Works Assoc.* 90(9): 112–21.

Lipp, E.K., J.L. Jarrell, D.W. Griffin et al. 2002. Preliminary evidence for human fecal contamination in corals of the Florida Keys, USA. *Mar. Pollut. Bull.* 44(7): 666–70.

Louis, V.R., E. Russek-Cohen, N. Choopun et al. 2003. Predictability of *Vibrio cholerae* in Chesapeake Bay. *Appl. Environ. Microbiol.* 69(5): 2773–85.

Lovins, W.A., J.S. Taylor, R. Kozik et al. 1999. Multi-contaminant removal by integrated membrane systems. Presented at the American Water Works Association Water Quality Technology Conference, Tampa, Florida.

MacAree, B.A., J.L. Clancy, G.L. O'Neill et al. 2005. Characterization of the bacterial population in RO distribution systems and their ability to form biofilms on pipe surfaces. Presented at the American Water Works Association Water Quality Technology Conference, Quebec City, Quebec.

Nasser, A.M., N. Zaruk, L. Tenenbaum et al. 2003. Comparative survival of *Cryptosporidium*, coxsackievirus A9 and *Escherichia coli* in stream, brackish and sea waters. *Water Sci. Technol.* 47(3): 91–6.

Paul, D.H. 1991. Osmosis: Scaling, fouling and chemical attack. *Desalination and Water Reuse* 1: 8–11.

Paul, J.H., J.B. Rose, J. Brown et al. 1995a. Viral tracer studies indicate contamination of marine waters by sewage disposal practices in Key Largo, Florida. *Appl. Environ. Microbiol.* 61(6): 2230–34.

Paul, J.H., J.B. Rose, S. Jiang et al. 1995b. Occurrence of fecal indicator bacteria in surface waters and the subsurface aquifer in Key Largo, Florida. *Appl. Environ. Microbiol.* 61(6): 2235–41.

Paul, J.H., J.B. Rose, S.C. Jiang et al. 1997. Evidence for groundwater and surface marine water contamination by waste disposal wells in the Florida Keys. *Water Res.* 31(6): 1448–54.

Paul, J.H., M.R. McLaughlin, D.W. Griffin et al. 2000. Rapid movement of wastewater from onsite disposal systems into surface waters in the Lower Florida Keys. *Estuaries* 23(5): 662–8.

Prévost, M., P. Laurent, P. Servais et al. (eds.). 2005. *Biodegradable organic matter in drinking water treatment and distribution.* Denver: American Water Works Association.

Reeves, R.L., S.B. Grant, R.D. Morse et al. 2004. Scaling and management of fecal indicator bacteria in runoff from a coastal urban watershed in Southern California. *Environ. Sci. Technol.* 38(9): 2637–48.

Ridgway, H.F., and H.C. Flemming. 1996. Membrane biofouling. In *Water treatment and membrane processes*, ed. J. Malleviale, P.E. Odendaal and M.R. Wiesner, 6.1–6.62. New York: McGraw-Hill.

Sinton, L.W., R.K. Finlay and P.A. Lynch. 1999. Sunlight inactivation of fecal bacteriophages and bacteria in sewage-polluted seawater. *Appl. Environ. Microbiol.* 65(8): 3605–13.

Tamburrini, A., and E. Pozio. 1999. Long-term survival of *Cryptosporidium parvum* oocysts in seawater and in experimentally infected mussels (*Mytilus galloprovincialis*). *Int. J. Parasitol.* 29(5): 711–5.

Taylor, J.S., and E.P. Jacobs. 1996. Reverse osmosis and nanofiltration. In: *Water treatment and membrane processes*, ed. J. Mallevialle, P.E. Odendaal and M.R. Wiesner, 9.18–9.22. New York: McGraw-Hill.

Taylor, J.S., J.D. Dietz, A.A. Randall et al. 2006. *Effects of blending on distribution system water quality*. Denver: American Water Works Association Research Foundation (Report 91065F).

USBR 2003. *Desalting handbook for planners*. 3rd ed. Denver: U.S. Department of the Interior, Bureau of Reclamation, Water Treatment Engineering and Research Group (Desalination and Water Purification Research and Development Report No. 72).

USEPA 1991. *Guidance manual for compliance with the filtration and disinfection requirements for public water systems using surface water sources*. Washington: U.S. Environmental Protection Agency (EPA No. 570391001). http://www.epa.gov/safewater/mdbp/guidsws.pdf.

USEPA. 1999. *Alternative disinfectants and oxidants manual*. Washington: U.S. Environmental Protection Agency, Office of Water (EPA 815-R-99-014). http://www.epa.gov/safewater/mdbp/alternative_disinfectants_guidance.pdf.

van der Hoek, J.P., J.A.M.H. Hofman, P.A.C. Bonne et al. 2000. RO treatment: Selection of a pretreatment scheme based on fouling characteristics and operating conditions based on environmental impact. *Desalination* 127(1): 89–101.

Vrouwenvelder, J.S., and D. van der Kooij. 2001. Diagnosis, prediction and prevention of biofouling of NF and RO membranes. *Desalination* 139(1–3): 65–71.

Wait, D.A., and M.D. Sobsey. 2001. Comparative survival of enteric viruses and bacteria in Atlantic Ocean seawater. *Water Sci. Technol.* 43(12): 139–42.

WHO. 2003. *Guidelines for safe recreational water environments. Vol. 1. Coastal and fresh waters*. Geneva: World Health Organization. http://www.who.int/water_sanitation_health/bathing/srwe1/en/.

WHO. 2008. *Guidelines for drinking-water quality*. 3rd ed. incorporating first and second addenda. Geneva: World Health Organization. http://www.who.int/water_sanitation_health/dwq/gdwq3rev/en/index.html.

WHO/WPC. 2006. *Health aspects of plumbing*. Geneva: World Health Organization and World Plumbing Council. http://www.who.int/water_sanitation_health/publications/plumbinghealthasp/en/index.html.

Yang, Z.B., and I.J. Hodgkiss. 2004. Hong Kong's worst "red tide"—causative factors reflected in a phytoplankton study at Port Shelter station in 1998. *Harmful Algae* 3(2): 149–61.

5 Monitoring, Surveillance, and Regulation

David Cunliffe, Shoichi Kunikane, Richard Sakaji, and Nick Carter

CONTENTS

The most effective means of ensuring drinking water safety is through the application of a comprehensive risk assessment and risk management approach. In the World Health Organization's (WHO) *Guidelines for Drinking-Water Quality* (GDWQ) (WHO 2008), this approach is based on application of a water safety plan (WSP) within a framework for safe drinking water. The framework includes (1) health-based targets, which provide the "benchmarks" for water suppliers; (2) the three components of WSPs: system assessment, operational monitoring, and management plans, which in combination describe the actions undertaken by water suppliers to ensure that safety, as defined by the health-based targets, is achieved; and (3) independent surveillance to assess the effectiveness of WSPs in meeting the targets. This chapter deals with the monitoring requirements to demonstrate that WSPs are appropriately designed, function effectively, and produce water that is safe to drink. The chapter also deals with independent surveillance and regulation of desalinated water supplies.

5.1 MONITORING

Monitoring plays a key role in ensuring that the WSP functions as intended and achieves the provision of safe drinking water. It includes validation, operational monitoring, verification, and surveillance. Validation is the process of obtaining evidence that control measures and the WSP as a whole are capable of achieving the health-based targets; operational monitoring is used to determine that individual components of a drinking water system are working as intended; verification provides assurance that the system as a whole is providing safe water; and surveillance reviews compliance with identified guidelines, standards, and regulations.

5.1.1 Validation

Validation of control measures, including treatment processes, is necessary to demonstrate that they are capable of operating as required. It is an investigative activity undertaken when a system is designed and constructed or modified. Validation should assess performance specifications for each control measure, taking into account source water characteristics. The first step of validation is to consider existing data available from the scientific literature, industrial bodies, other users of the equipment, and manufacturers or suppliers. As identified in Chapter 2, a body of evidence is available on the performance of a range of standard processes used to desalinate water. Validation should confirm that specific unit processes achieve accepted performance standards. Monitoring associated with validation is normally an intensive activity undertaken typically in pilot trials or precommissioning and commissioning of systems. Monitoring requirements will generally be more extensive for nonstandard or innovative processes. Validation can assist in identifying operating modes, operational monitoring parameters, and criteria as well as maintenance requirements.

5.1.2 Operational Monitoring

Operational monitoring is the planned series of observations or measurements undertaken to assess the ongoing performance of individual control measures in preventing, eliminating, or reducing hazards. It will normally be based on simple and rapid procedures such as measurement of turbidity and chlorine residuals or inspection of distribution system integrity.

Parameters used for operational monitoring should

- Be readily measured
- Reflect the effectiveness of each control measure
- Provide a timely indication of effectiveness based on compliance with operational limits
- Provide an opportunity for implementation of corrective measures where required

Operational limits separate acceptable from unacceptable performance of control measures. If unacceptable performance is detected, then predetermined corrective actions need to be taken. The aim is that measurement of operational parameters and implementation of any required corrective action should be undertaken within a time frame that prevents unsafe water from being supplied to consumers. In some cases, operational target criteria that are more stringent than operational limits can be used to provide early warnings to operators to enable changes to be implemented prior to limits being reached or exceeded.

5.1.3 Verification

In addition to operational monitoring of individual components of a drinking water system, it is necessary to undertake verification to provide assurance that the system as a whole is operating safely. The range of parameters included in verification will be directed by established drinking water standards and guidelines and will typically include testing for indicators of microbial quality as well as chemical hazards. Verification can be undertaken by the water supplier as part of quality control, by an independent surveillance agency or by a combination of these two.

5.1.4 Surveillance

Surveillance is the "continuous and vigilant public health assessment and review of the safety and acceptability of drinking water supplies" (WHO 1976). Surveillance should be undertaken by an independent agency that is provided with legislative powers to support its activities and enforcement of any corrective actions required to protect public health.

There are two approaches to surveillance: approaches based on direct assessment and auditing. In the audit-based approach, testing of drinking water quality, as supplied to consumers, is undertaken by the water supplier with assessment of monitoring programs, and results are included in the auditing process. In the direct assessment approach, independent testing of water quality is undertaken by the surveillance agency. This type of testing complements the verification program undertaken by the supplier. Surveillance will normally also include processes for the approval and review of the WSP as well as a review of the implementation of the WSP.

5.2 OPERATIONAL MONITORING FOR DESALINATION

Hazards can occur at any step from the source of seawater or brackish water to the final drinking water tap in a community. Control measures are used to prevent, eliminate, or reduce these hazards and can include actions such as the protection of source waters from contamination by sewage or industrial wastes, treatment, and protection of distribution systems. Chapter 2 provides a discussion of typical technologies and control measures that can be applied to produce safe drinking water through desalination. The combination of control measures used in particular schemes will vary depending on a range of factors, including the type of source water and size of the scheme. Irrespective of the combination of control measures used in a scheme, the effectiveness of the performance of each measure needs to be assessed. This is achieved by operational monitoring.

Operational monitoring can be applied to each of the components of a typical desalination treatment and distribution process, as described in Chapter 2:

- Source water, taking into account potential impacts of contamination sources and fluctuations caused by natural events
- Pretreatment, including the use of additives and chemicals
- Desalination (membrane-based and thermal processes)
- Blending (including treatment of blending water), remineralization, and disinfection
- Storage and distribution, including an assessment of the corrosivity and stability of the product/blended water

Ensuring consistent drinking water quality requires care and attention to details in the day-to-day operation of all processes within a desalination system, from source to consumer. Such details should be reflected in the operational management of the desalination facility. This should be manifested in an accurate and up-to-date operations plan and should be documented and supported by appropriate training for plant operators.

Table 5.1 provides a summary of operational monitoring parameters with suggested monitoring frequencies for large plants (>38,000 m^3/day) and small plants (<38,000 m^3/day). Where resources are limited, the selection of operational parameters and frequencies should be guided by a risk assessment in the WSP. Using this approach, monitoring of hazards and control measures can be prioritized on the basis of the level of risk associated with the hazard or with the loss of effective

TABLE 5.1
Suggested Monitoring Parameters and Frequencies for Desalination Plants[a]

Component	Control Measures	Operational Parameters	Monitoring Frequency[b,c]	
			Large Plant	Small Plant
Source water	Detect and prevent contamination by sewage (pathogenic protozoa, viruses, bacteria) (likelihood of presence based on sanitary inspection)	Enterococci and/or *E. coli*	D or W	W
	Detect and prevent impacts of storm events	Turbidity (used as online measurement for process control)	Preferably online	Preferably online
	Detect and prevent impacts of microalgae or cyanobacteria	Algal species, including cyanobacteria, dinoflagellates, or chlorophyll as a surrogate	W or M	M
	Detect and prevent impacts by industrial discharges (based on an assessment of local conditions)	TOC (if concentrations change, investigate sources)	W	M
		Petroleum oil hydrocarbons/grease, including volatile compounds	W	Y
		Industrial chemicals	W	M
		Radioactivity	Y	Y
	Monitoring associated with downstream control measures (pretreatment and treatment)	Salinity	D	D
		Chloride	D	D
		Sodium	W	W
		Boron	M	Y
		Bromide	M	Y
		Silica	D	D
		Iron	D	W
		Manganese	M	Y
		Turbidity	Online	Online
		Alkalinity	D	D
		pH	Online	Online
		Temperature	Online	Online
		Heavy metals	W	M

(continued on next page)

TABLE 5.1 (continued)
Suggested Monitoring Parameters and Frequencies for Desalination Plants[a]

Component	Control Measures	Operational Parameters	Monitoring Frequency[b,c]	
			Large Plant	Small Plant
		Low-solubility chemicals (e.g., calcium, fluoride, barium, strontium, magnesium, sulfate)	W	M
		Hydrogen sulfide and metal sulfides	W	M
		Ammonia	W	M
		TDS	D	D
Pretreatment: Membranes	Detection and prevention of biofouling, scaling, precipitation	SDI	Online	Online
		Flow rates	Online	Online
		Conductivity	Online	Online
		Conductivity/TDS ratios	D	D
		Turbidity after pretreatment, particle counts	Online	D
		pH (if acidification or alkalinization)	Online	Online
	Use of additives (e.g., antiscalant)	Flow and dose rate monitoring	Online	Online
	Quality control on additives and materials	Test additives and materials; check records	D	D
	Prevention of microbial fouling	Disinfectant residual or ORP	Online	Online
Pretreatment: Thermal processes	Use of additives (e.g., antiscalant, antifoaming)	Flow and dose rate monitoring	Online	Online
		pH (if acidification used)	Online	Online
	Quality control on additives and materials	Test additives and materials; check records	D	D
	Prevention of microbial fouling	Disinfectant residuals	Online	Online
Process management: Membranes		Recovery ratio (calculated from flow rates)	D	D

TABLE 5.1 (continued)
Suggested Monitoring Parameters and Frequencies for Desalination Plants[a]

Component	Control Measures	Operational Parameters	Monitoring Frequency[b,c]	
			Large Plant	Small Plant
		Chemical balance from conductivities and flow rates (calculated)	W	M
		Transmembrane pressure	Online	Online
		Flow meters on permeate and brine	Online	Online
		Conductivity in permeate and brine	Online	Online
		TOC (particularly where source water contains elevated microbial contamination)	Online	Online
Process management: Thermal processes		Incoming steam pressure and temperature	Online	Online
		Makeup and distillate flow rate	Online	Online
		Distillate and condensate conductivity	Online	Online
		TBT (MSF/MED)	Online	Online
		Copper, iron, and nickel (corrosion)	W	Q
		Hydrazine and other boiler treatment chemicals (MED-TC) (depending on design)	W	M
Blending	Preventing microbial contamination (use only pretreated water or protected groundwater; untreated surface water not to be used)	*E.coli*/enterococci in blending water	D	W
		Appropriate parameters for processes used to treat blending water (e.g., turbidity for filtration, disinfectant dose/concentration)	Online	Online

(continued on next page)

TABLE 5.1 (continued)
Suggested Monitoring Parameters and Frequencies for Desalination Plants[a]

Component	Control Measures	Operational Parameters	Monitoring Frequency[b,c]	
			Large Plant	Small Plant
	Providing chemical stability, maintaining minimum calcium and magnesium concentrations	LSI/CCPP where mortar linings used	D	D
		LR where steel or carbon/steel used	D	D
		Calcium and magnesium	D	D
		pH	Online	Online
		Conductivity	Online	Online
Remineralization	Providing chemical stability, maintaining minimum calcium and magnesium concentrations	LSI/CCPP where mortar linings used	D	D
		LR where steel or carbon/steel used	D	D
		Calcium and magnesium (or total hardness)	D	D
		Alkalinity	D	D
		pH	Online	Online
		Conductivity	Online	Online
	Quality control on additives and materials	Test additives and materials; check records	D	D
Disinfection	Removal of microbial contaminants	Disinfectant dose monitoring	Online	Online
		Calculate CT	D	W
		HPC	M	Y
		DBPs (including brominated compounds) relevant to the method of disinfection	W	M
Corrosion inhibition	Reduction of corrosion in distribution systems using inhibitors such as phosphates and silicates	Flow and dose rate monitoring	Online	Online
	Quality control on additives and materials	Test additives and materials; check records	D	D

TABLE 5.1 (continued)
Suggested Monitoring Parameters and Frequencies for Desalination Plants[a]

Component	Control Measures	Operational Parameters	Monitoring Frequency[b,c]	
			Large Plant	Small Plant
Storage distribution	Preventing microbial and chemical contamination by controlling intrusion through cross-connections/backflow or faults in mains or other infrastructure	*E. coli*	D	W
		HPC	D	M
		Turbidity	D	D
		Inspect for system integrity, monitor burst main frequency and repairs	W	M
		Monitor system leakage	M	Y
	Control of free-living microorganisms	Disinfectant residual (consider persistent disinfectant where *Legionella/Naegleria* potential considered unacceptable)	D	D
	Prevention of corrosion in storage tanks, long pipes, domestic plumbing	pH	D	D
		Iron	M	M
		Zinc	M	Y
		Nickel	M	Y
		Copper	M	Y
		Lead (if problem)	M	M
		Zinc and phosphate (if corrosion inhibitors are used)	W	W
	Maintain chemical stability after mixing different sources (desalinated/ nondesalinated water) or after disinfection	Postmixing or postdisinfection monitoring for LSI/ CCPP where mortar linings used or LR where steel or carbon/ steel used	D	D
	Disinfection	DBPs (including brominated compounds)	W	M
Concentrate discharges	In marine or brackish lake environments, select discharge points to minimize impacts; discharge into areas with high levels of mixing or use diffusers to promote mixing	Temperature and dissolved oxygen (thermal processes)	Online	Online
		pH	Online	Online
		Salinity	Online	Online
		Heavy metals/salts	M	Q
		Additives	M	Q
		Phosphates and nitrates	M	Q

(continued on next page)

TABLE 5.1 (continued)
Suggested Monitoring Parameters and Frequencies for Desalination Plants[a]

Component	Control Measures	Operational Parameters	Monitoring Frequency[b,c]	
			Large Plant	Small Plant
	For brackish aquifer, apply discharge requirements set by environmental protection agencies	Temperature (thermal processes)	Online	Online
		pH	Online	Online
		Salinity	Online	Online
		Heavy metals/salts	M	Q
		Additives	M	Q
		Phosphates and nitrates	M	Q
Cooling water discharges	In marine or brackish lake environments, select discharge points to minimize impacts; discharge into areas with high levels of mixing or use diffusers to promote mixing	Temperature	Online	Online
		Dissolved oxygen	Online	Online
		Corrosion products (copper, iron, nickel)	W	Q
		Disinfectant residuals	D	W
	For brackish aquifer, apply discharge requirements set by environmental protection authorities	Temperature	Online	Online
		Dissolved oxygen	Online	Online
		Corrosion products (copper, iron, nickel)	W	Q
		Disinfectant residuals	D	W
Wastewater effluents from pretreatment facilities or from membrane cleaning	Collect discharges and treat or discharge in accord with requirements set by environmental protection agencies	pH	Online	Online
		Turbidity	Online	Online
		Suspended solids	D	D
		Residual disinfectants	D	D
		Iron or aluminum (based on the type of coagulant used)	W	M
		Membrane-cleaning agents	On discharge	On discharge

[a] These suggestions are intended to provide perspective for decision makers to determine the most appropriate monitoring type, frequency, and locations to suit their particular circumstances and needs.

[b] Monitoring frequency: D, daily; W, weekly; M, monthly; Q, quarterly; Y, yearly. In some cases, monitoring requirements will be based on the source of water (seawater, estuarine, or groundwater) and sanitary surveys. For example, the frequency of petroleum hydrocarbon monitoring will be influenced by the likelihood of potential sources of contamination.

[c] The capacity of a "large" or "small" plant is assumed to be more or less than about 38,000 m^3/day, respectively.

performance of the control measure designed to reduce or remove the hazard. The frequency of testing should also be based on the level of risk in conjunction with a consideration of variability in the presence of the hazard or in the performance of the control measure.

5.2.1 Source Water

Desalination plants can treat seawater (from offshore intakes and pipelines or from wells located on the seabed, beach, or inland in coastal areas and on islands) and brackish surface water or groundwater. As indicated in Chapter 2, desalination facilities require reliable sources of water with a reasonably consistent quality. The quality and consistent nature of intake water will influence the nature and effectiveness of pretreatment processes, the performance of treatment processes, and the safety of drinking water produced by desalination. Quality is affected by naturally occurring elements, by events such as storms and spills, and by ongoing sources of contamination. Wherever possible, a preventive approach should be applied to ensure the quality of the source water, and priority should be given to control measures designed to protect source waters and intake pipes. The less desirable alternative is to design treatment processes to remove preventable contaminants. In this case, poor performance or short-term failure of downstream treatment processes can have greater consequences. In addition, the presence of contaminants may be hazardous if source water is used in blending processes.

5.2.1.1 Marine Waters

Sources of contamination with potential impacts on water quality and desalination plant intakes can include the following:

- Domestic wastewater discharged raw or partially treated (predominantly microbial pathogens and some persistent chemicals)
- Onshore and offshore dumping of hazardous wastes (chemicals, pathogens, and radioactive materials)
- Offshore and nearshore oil and gas exploration, excavation, production, and refining processes (drilling mud turbidity; chemicals, including hydrocarbons, small doses of radionuclides)
- Brine and other waste streams from desalination plants (chemicals, including antiscalants, antifoaming agents)
- Transport cargo, including oil tankers and tourist vessels (ballast water, hydrocarbon spills, wastewater containing microbial pathogens)
- Discharges from industrial complexes (chemicals, including hydrocarbons and heavy metal–organic compound complexes)
- Effluents from power generation plants, including cooling waters (antifouling agents, biocides, and thermal pollution)
- Military actions where oil spills or chemical, biological, and possibly radioactive pollutants are produced accidentally or intentionally

- Eutrophication associated with discharge of nutrient-rich storm waters
- Increased organic loadings associated with fish kills or decomposing marine life
- Trapping and blockage of intake screens by aquatic organisms or passage of organisms through screens and into process equipment. Organisms such as mussels, clams, and mollusks can grow in intake structures, whereas microorganisms can attach and grow, producing biofilms

5.2.1.2 Brackish Surface Water or Groundwater

Brackish water can be withdrawn from surface sources such as lakes and estuaries or from aquifers. Open lakes and estuaries can be subject to sources of pollution similar to those affecting seawater. Groundwater, particularly when taken from deep or confined aquifers, is generally more consistent in quality and contains lower levels of contamination than surface water as a result of the filtering effects of soil barriers. However, contamination of groundwater can occur and is of growing concern. Sources can include

- Domestic wastewater discharged raw or partially treated (predominantly microbial pathogens as well as detergents and household cleaning and disinfecting chemicals)
- Industrial discharges (chemicals, including hydrocarbons)
- Hazardous waste dumps (chemicals)
- Soluble fertilizers and pesticides from agriculture
- Oil exploration activities, oil industry products, wastes, and heavy petroleum oil derivatives

5.2.1.3 Operational Monitoring Parameters

Operational monitoring will be influenced by investigations and knowledge of source water characteristics, potential sources of contamination, and the location of feedwater intakes in relation to these sources. Investigations undertaken as part of intensive precommissioning surveys or following commissioning will determine the range of parameters and sampling frequencies included in operational monitoring programs. All monitoring needs to have a purpose and to incorporate mechanisms for interpretation and action. Operational limits should be established for parameters selected for monitoring. If these limits are exceeded, causes should be investigated and remedial action initiated. Wherever possible, remedial measures should be identified and documented prior to limits being exceeded.

Enterococci and/or *Escherichia coli* can be used as indicators of bacteria from sewage contamination, whereas chemical parameters can include ammonia (sewage contamination); oil hydrocarbons, including volatile compounds and greases; industrial chemicals (parameters industry dependent), and radioactivity. Total organic carbon (TOC) changes could be used as a general indicator of contamination by organics or sewage, with changes in concentration leading to investigations of potential causes.

Storm events can lead to deterioration in water quality. Impacts of storm events can be assessed by monitoring turbidity; if operational limits are exceeded, one option can be to shut down intakes until turbidities fall to normal levels.

Microalgal blooms may also be a cause of increased turbidity or a source of algal toxins. In surface waters prone to blooms, algal species, including dinoflagellates and cyanobacteria, or chlorophyll should be monitored. Other organisms, including seaweeds, could cause blockages of intake structures and may also need to be monitored. The occurrence or abundance of organisms may be seasonal, and monitoring programs should take this into account. Impacts of organisms on intake structures, including blockage of screens and growth within intake pipes, should be monitored. Biofilms and biofouling of intake structures may be controlled using disinfectants as part of pretreatment processes.

Depending on source water characteristics, there is a range of parameters that can have potential downstream impacts on treatment processes and hence influence pretreatment requirements. These parameters should be assessed and considered for inclusion in monitoring programs. Parameters could include

- Heavy metals
- Low-solubility chemicals, such as calcium carbonate, calcium sulfate, calcium fluoride, barium sulfate, strontium sulfate, magnesium hydroxide
- Turbidity, alkalinity, and ph
- Silica
- Hydrogen sulfide and metal sulfides (particularly in groundwater)
- Boron and bromide (particularly in seawater)
- Iron, manganese, and alumina (in groundwater)
- Total dissolved solids (TDS)
- Disinfectants and other treatment chemicals in cooling water discharges from power plants

Temperature and pH should be monitored because of their impacts on pretreatment and treatment processes. The temperature of groundwater remains relatively constant, but surface water temperatures can vary widely even within the same day and should be monitored more frequently. Measurement of pH and alkalinity is important in relation to control of corrosion and efficiency of the coagulation process used as pretreatment for membrane-based processes.

5.2.2 Pretreatment

Pretreatment can incorporate the addition of chemicals. Tables 2.1 and 2.2 in Chapter 2 present typical chemicals used in pretreatment before thermal and membrane processes. Chemicals can be used to facilitate the performance of pretreatment processes or as cleaning agents (see Table 2.3). In addition to ensuring that additives and chemicals produce the required result, care also needs to be taken to ensure that they are of sufficient quality, do not contain undesirable contaminants, and are not toxic. Quality control of chemicals is discussed in Section 5.5.

5.2.2.1 Membrane Processes: Scaling, Precipitation, and Fouling

Pretreatment is a requirement for membrane-based processes to reduce scaling, precipitation, and fouling. The main causes are

- Particulates and suspended solids
- The presence of low-solubility salts, such as calcium carbonate, calcium sulfate, calcium fluoride, barium sulfate, strontium sulfate, magnesium
- Organic foulants (oil and grease, hydrocarbons)
- Colloidal silica and sulfur
- Metal oxides (iron, manganese, and alumina)
- Overfeed of antiscalants
- Biofouling caused by biological growths

One aim of pretreatment is to remove suspended solids and turbidity in order to achieve a required silt density index (SDI) value. Coagulation (with pH adjustment) and filtration are the most common forms of pretreatment used for seawater; for brackish water, pretreatment may also include settling and softening. Desalination feedwater should have a turbidity below 1 nephelometric turbidity unit and an SDI below 5 to ensure reliable performance. Turbidity (or, alternatively, particle counts) can be monitored online. SDI can be determined on-site, with the frequency of monitoring depending on variability and system characteristics. A second aim is to reduce downstream impacts on treatment processes such as scaling, precipitation, and fouling. Scale inhibitors (antiscalants) are used to reduce the impacts of low-solubility salts or specific chemicals such as silica, whereas disinfectants are used to reduce the likelihood of biofouling.

Membranes can be damaged by oxidizing agents such as free chlorine, which need to be removed from feedwater flows. Dechlorination is generally achieved using sodium bisulfite. The performance of chlorination and dechlorination processes can be measured online using residual analyzers or, alternatively, oxidation–reduction potential (ORP) or redox potential monitors. In addition, dosing should be controlled by monitoring injection rates and levels in chemical tanks. Quantities of chemicals used and average dose rates should be checked and recorded on a routine basis by plant operators.

5.2.2.2 Thermal Processes: Multistage Flash Distillation and Multiple Effect Distillation

Thermal processes generally require less pretreatment than membrane-based processes. The primary issues are biofouling of intake pipes, corrosion, and scaling. Corrosion and scaling can be minimized by a combination of pretreatment and design features of the downstream process. Corrosion is primarily caused by dissolved gases, and physical deaeration is needed for multistage flash distillation (MSF) plants to remove oxygen. In multiple effect distillation (MED) plants, deaeration takes place in the spraying nozzles. Scaling can be caused by the precipitation of calcium sulfate, calcium carbonate, or magnesium hydroxide. Scale inhibitors can be added during pretreatment, or, alternatively, acid can be used to lower the pH and prevent the production of scale. Owing to the improvement of antiscalant products, acid treatment is generally not required for seawater desalination and is used only in special cases (brackish waters or wastewaters with high concentrations of carbonic species).

Biological growth can occur in feedwater intakes and supply lines. Chlorination can control this growth. Free chlorine concentrations should be maintained at about 0.1 mg/L. Periodic shock injections at higher doses may also be required, depending on source water characteristics.

In addition to antiscalants and disinfectants, antifoaming agents are used in thermal processes to disperse foam-causing organics and to reduce surface tension at the steam–water interface. If scale inhibitors, disinfectants, or antifoaming agents are used, the quality of chemicals needs to be carefully checked before use. In addition, dosing should be controlled by monitoring injection rates and levels in chemical feed tanks. Quantities of chemicals used and average dose rates should be measured and recorded on a routine basis by plant operators.

5.2.3 Treatment

5.2.3.1 Membrane Processes

The effectiveness of membrane processes can be assessed by online monitoring of transmembrane pressures, longitudinal pressure drops, and differential pressures. In addition, conductivities and flow rates of permeate and concentrate streams should be monitored and used to determine recovery ratios and chemical balances. Scaling, precipitation, and fouling due to inadequate pretreatment can cause a gradual decline of longitudinal differential pressures, transmembrane pressure or permeate flux, and an increase in salt permeation with time. These consequences can be detected by operational monitoring of membrane treatment processes. Changes in conductivity/TDS ratios in membrane product water can also be an indicator of scaling, precipitation, and fouling.

Monitoring membrane pressures and permeate conductivities should give an indication of membrane integrity, but minor damage allowing for passage of microbial contaminants may not be detected. The sensitivity of conductivity measurement is limited. TOC monitoring has been suggested as an alternative, particularly when source water contains elevated levels of microbial contamination. However, there are also questions about the sensitivity of this parameter, and further research is required to identify sensitive integrity-testing procedures.

Temperatures should also be included in operational monitoring plans. A temperature probe could be installed prior to the membranes. High temperatures may decrease the energy requirement of the reverse osmosis (RO) unit and improve water fluxes but accelerate the increase of the salt passage index over time.

5.2.3.2 Thermal Processes

The choice of the top brine temperature (TBT) is a key design feature of thermal processes, and it should be continuously monitored. Corrective action and possibly shutdown of the plant should occur if the TBT exceeds a predetermined target value. Calcium sulfate and calcium carbonate scaling can be controlled through operating thermal processes at controlled temperatures. This is achieved at the design stage by the choice of the TBT and of other parameters, such as the concentration of brine. For example, the maximum TBT of an MSF plant should not be above 112°C, whereas the maximum for an MED plant should be limited to 65°C.

Operational monitoring should also include the following:

- TDS can be continuously monitored using conductivity meters. When concentrations exceed target criteria, distillate should be rejected. Should the distillate conductivity become high, the cause should be investigated (e.g., leakage of a tube or displacement of a demister).
- Distillate and makeup flow rates should also be monitored online. If the makeup flow rate falls below a given value, corrective action should be taken, and possibly the plant should be shut down.
- Systems are manufactured from a range of materials, including steel and alloys of copper and nickel. Copper, iron, and nickel can be measured in distillate to assess corrosion and equipment longevity.
- Operating temperatures should be monitored.

In some MED processes incorporating a thermocompressor device (MED-TC), steam condensate may mix with distillate. In this case, boiler treatment chemicals, including hydrazine, carbohydrazide, and other oxygen scavengers (used for treatment of high-pressure boilers in power plants), could be carried over into distillate. Ideally, this issue should be dealt with in the design stage. However, monitoring may need to be considered.

5.2.4 Blending and Remineralization

Blending is undertaken to stabilize desalinated water, reduce corrosiveness, and improve the taste and acceptability of water supplied to consumers. Two types of approaches are taken to achieve these aims: (1) blending with higher-salinity water and (2) remineralization through the addition of chemicals. Blending requires the proper design of mixing facilities to overcome differences in the physical properties of the fluids being blended. Adequate energy must be supplied to overcome density differences in the two fluids and provide dispersion of the fluid elements from the different streams to produce a homogeneous mixture. The blended product should be reliably produced and especially of a consistent and safe microbiological quality.

The choice of blending water is important from both chemical and microbiological perspectives. From a chemical standpoint, the blending water should produce a product that is stable and will not act aggressively toward the distribution system. Well-operated membrane-based and thermal distillation processes should provide desalinated water that is free from pathogens. Blending waters should be of a similar quality. For this reason, blending should generally be achieved using pretreated source water. Alternatives are protected groundwater or treated surface water. As discussed in Chapter 4, treatment of blending waters will generally be required unless it is microbiologically safe (e.g., protected groundwater).

When blending waters of disparate quality, potential chemical reactions and chemical solubility issues should be considered. For example, if blending leads to calcium carbonate scaling, there could be adverse impacts on blending performance, leading to fluctuations in product water quality. Scaling can foul static mixing blades, causing short-circuiting in the blending unit and leading to pockets of highly saline water in the product. The potential for such problems occurring can be assessed by operational monitoring for indices such as the Langelier saturation index (LSI), the calcium carbonate precipitation potential (CCPP), or the Larson ratio (LR), as discussed in Chapter 2, and through periodic inspections of blending systems.

Posttreatment may be achieved by the addition of chemicals as described in Chapters 2, 3, and 4. If this is undertaken, there are three primary concerns that need to be addressed:

1. The quality of the additives and the introduction of chemical contaminants produced during manufacturing, storage, distribution, and transport (see Section 5.5). Unlike pretreatment chemicals, there are no downstream processes that will remove undesirable contaminants.
2. Controlling dose rates to ensure that required concentrations are provided.
3. Preventing or minimizing unwanted chemical reactions following chemical addition. This issue is similar to blending. Localized changes can occur at dosing points, leading to fouling problems on a microscale.

Operational parameters for posttreatment will vary depending on the process, the potential impacts on desalinated water quality, and the nature of the distribution network. Online measurement of pH and conductivity can be used to monitor posttreatment processes involving either blending or chemical addition. Calcium, magnesium, and alkalinity should also be monitored to ensure that minimum concentrations are achieved. Dosing should be controlled by monitoring injection rates and levels in chemical tanks. Quantities of chemicals used and average dose rates should be checked on a routine basis by plant operators.

In terms of corrosivity, indices such as the LSI or CCPP can be used to ensure that the desalinated water will have minimal corrosion impact on mortar-lined materials; if steel or carbon/steel materials are present in the distribution system, the LR should be used in addition to the LSI or CCPP.

As discussed in Chapter 4, depending on the source of blending waters, *E. coli*, enterococci, or coliphage can be used to assess potential impacts of fecal pollution. Processes used to treat blending waters should be monitored to ensure that they are effective.

5.2.5 Posttreatment Disinfection

General principles of posttreatment disinfection of desalinated water are similar to those of disinfection of freshwater sources of drinking water. However, the presence of higher concentrations of bromides from seawater and other brackish water sources may lead to the formation of brominated disinfection by-products (DBPs). The presence of bromide can give rise to bromates if ozone is used, and bromide concentrations above 0.4 mg/L can reduce the stability of chloramines if chloramination is used to disinfect desalinated water.

As for other posttreatment chemicals, dosing needs to be monitored. Dosing should be controlled by monitoring injection rates and residual levels in chemical tanks. Quantities of chemicals used and average dose rates should be checked on a routine basis by plant operators. Disinfection concentrations or doses can be monitored using online devices or manually using field kits. Where disinfection is applied using chlorine or chloramines, residual concentrations can be measured. In combination with flow rates and detention times prior to supply to consumers, the residuals can be used to calculate disinfectant concentration × time (CT) values (see Chapter 4). Where ultraviolet light irradiation is used, minimum doses can be monitored using online devices.

Concentrations of DBPs should be monitored. The composition of by-products will depend on the nature of disinfection: for example, trihalomethanes (THMs) for chlorination, THMs and cyanogen chloride for chloramination, bromate and brominated THMs for ozonation, and chlorite and chlorate for chlorine dioxide.

5.2.6 Storage and Distribution

A point of differentiation between desalinated water supplies and other types of drinking water supply is the potential for increased corrosion unless the water is appropriately stabilized by posttreatment blending or addition of chemicals. Posttreatment monitoring should be extended into the distribution system to assess the impacts of desalinated water on storage and distribution systems. Monitoring should include testing for corrosion products based on the types of materials used in distribution networks as well as in domestic plumbing. This could include testing for copper, nickel, iron, zinc, and lead. If monitoring detects evidence of elevated levels of corrosion, posttreatment blending processes should be reviewed. In some circumstances, the choice of materials used should also be reviewed.

Desalination systems are more prevalent in warmer climates, and elevated temperatures of water in storages and distribution systems may increase the potential for growth of free-living pathogens such as *Legionella* and *Naegleria fowleri* (see Chapter 4). In these climates, underground installation of pipelines will help control water temperatures and reduce risks from these organisms. Maintaining free or combined chlorine residuals in distribution systems can also provide protection. If this approach is adopted, operational monitoring should include testing for free chlorine residuals throughout the distribution system. An alternative strategy to provide protection from microbial growth in distribution systems is to achieve disinfection using chloramination rather than chlorination. In this case, operational monitoring should include testing for total chlorine residuals throughout the distribution system.

At the end of the distribution pipeline, problems can be introduced when homeowners, particularly in areas where water supply has historically been unreliable, install storage reservoirs, usually on the roof, which can have a negative effect on the quality of water that comes out of the user's tap because of the entry of contamination.

Other types of operational monitoring could be required for control measures that are common to desalinated and nondesalinated supplies. As shown in Table 5.1, these could include measures applied to minimize intrusion of microbial or chemical contamination in distribution systems through cross-connections or backflow. Further information is provided in the GDWQ (WHO 2008) and the supporting texts *Safe Piped Water: Managing Microbial Water Quality in Piped Distribution Systems* (Ainsworth 2004) and *Health Aspects of Plumbing* (WHO/WPC 2006).

5.3 DISCHARGES, INCLUDING CONCENTRATES, COOLING WATER, PRETREATMENT RESIDUALS, AND MEMBRANE-CLEANING SOLUTIONS

As described in Chapter 2, desalination can produce a number of discharges, including

- Residuals from pretreatment processes used prior to membrane-based desalination. The residuals can contain solids and sludge together with coagulants and residual disinfectants. These residuals can be treated or in some cases discharged without treatment.
- Membrane-cleaning solutions and subsequent flushing waters.
- Concentrates from membrane and thermal processes, including chemicals removed by the desalination process together with additives such as coagulants, scale inhibitors, and antifoaming agents. Thermal discharges may be at elevated temperatures and can contain low levels of dissolved oxygen.
- Cooling water discharges from thermal processes. These can be discharged separately or blended with brine concentrates. The temperature of cooling water discharges will typically be around 10°C above the temperature of the source waters, and dissolved oxygen concentrations may be reduced. Discharges may contain small amounts of corrosion products, disinfectants, and DBPs.

Concentrates and other process wastes may be discharged through purpose-built outfalls, existing outfalls associated with wastewater treatment plants or other industrial plants. In some cases, particularly for smaller plants, concentrates, residuals, and cleaning solutions may be discharged into wastewater systems.

The composition and volumes of discharges should be monitored to determine the chemical composition of the waste stream and to enable assessment of potential environmental impacts on receiving waters. Temperature and dissolved oxygen in discharges from thermal processes should be monitored. Where combined discharges are produced, the quality of the mixed effluents should be monitored. Discharges need to meet local, regional, or national requirements established by environmental protection agencies, and monitoring programs will need to be consistent with these requirements.

Development of bioassays could provide a useful tool to assess environmental impacts from the mixture of chemicals and physical properties in the various discharges produced during desalination. Where concentrates and residuals are discharged to wastewater treatment plants, monitoring should be undertaken to determine compatibility with the wastewater quality and with any associated wastewater recycling schemes. Relative flows of wastewater and desalination discharges need to be measured as an indicator of potential variations in water quality.

A comprehensive monitoring program should include the following:

- Pretreatment residuals should be monitored for turbidity/suspended solids, coagulant chemicals, residual disinfectants, and pH.
- Membrane-cleaning solutions should be monitored for cleaning chemicals.
- Brine discharges should be monitored for TDS, salts, heavy metals, nutrients, temperature, and dissolved oxygen (thermal processes) and additives, such as antiscalants and antifoaming agents.
- The temperature and dissolved oxygen of cooling water discharges should be monitored, together with copper, nickel, and iron as indicators of corrosion products.

5.4 VERIFICATION

Verification of desalinated water quality follows the same principles as those applied to other types of drinking water supply in the WSP (WHO 2008). The aim of verification is to ensure that desalinated water systems produce safe and acceptable drinking water that meets the health-based targets. The range of parameters included in verification monitoring will be directed by established drinking water standards and guidelines. Some verification could be performed by the desalinated water producer, and some by the water supplier in the distribution system.

Verification of microbial quality will typically include testing for fecal indicator organisms and could include testing for heterotrophic plate count (HPC) (see Chapter 4). Testing for specific pathogens is generally not justified, although there could be some exceptions. For example, in hot climates, testing could be included for organisms such as *Legionella* or *Naegleria fowleri*, which can grow within distribution systems. Verification of microbial quality needs to take into account all possible sources of contamination, including desalination product water and blending waters, as well as ingress during storage and distribution. Information on microbial quality in distribution systems is provided in the document *Safe Piped Water: Managing Microbial Water Quality in Piped Distribution Systems* (Ainsworth 2004). Verification programs for microbial quality will generally include a number of locations, starting at completion of treatment and including points within the distribution system. The number of samples will be a function of system complexity and size, and guidance is provided in the GDWQ (WHO 2008).

The location and frequency of testing for chemical parameters will depend on their potential variability and principal sources. Sampling at the end of treatment is generally sufficient for parameters that are not impacted by distribution; for parameters such as DBPs, where concentrations can change, sampling should include locations at the extremities of distribution systems. Because of the presence of bromide in seawater, DBPs are likely to be dominated by brominated compounds. This could also be an issue for brackish surface water and some groundwaters. Monitoring programs should include testing for the presence of chemicals used in treatment processes as well as for calcium, magnesium, and other mineral salts. Unless specific treatment (polishing) is applied, boron concentrations in seawater desalinated by reverse osmosis are likely to exceed the current WHO guideline value (WHO 2008) because of the limited removal efficiency of naturally occurring concentrations. However, as noted in Chapter 3, it is likely that the health-based guideline value will be increased in the fourth edition of the GDWQ.

Owing to the potentially aggressive nature of desalinated water, verification testing by the water supplier could include testing for corrosion products from distribution systems. This testing should include corrosion products arising from household plumbing, including nickel, lead, chromium (from plated parts and fixtures), and copper.

Further guidance on identifying priorities for chemical testing is provided in the GDWQ supporting text, *Chemical Safety of Drinking-Water: Assessing Priorities for Risk Management* (Thompson et al. 2007).

Acceptability is an important aspect of desalinated water supplies, with specific postdesalination measures undertaken to improve taste. Verification should include an assessment of acceptability to consumers. In addition to testing for calcium and magnesium, this can include regular testing for parameters such as TDS and pH as well as monitoring of consumer complaints and comments. Deviations from the norm are particularly noticeable to consumers, who may interpret aesthetic problems as indicating health risks. Although acceptability is largely a subjective judgment, complaints can sometimes provide valuable information leading to detection of problems that may not have been identified by routine monitoring.

5.5 QUALITY CONTROL

Quality control of chemicals, maintenance of monitoring equipment, review of laboratory performance, and selection of test methods are important components of monitoring programs. Chemicals and additives used in desalination processes need to be of sufficient quality to not contain undesirable concentrations of contaminants or be toxic. Quality control of activities associated with producing data is required to ensure that results are accurate and meaningful.

5.5.1 Additives and Chemicals

Raw materials, side reactions, and decomposition reactions during chemical manufacturing, production, or storage can create unwanted by-products in chemical feedstocks. In some cases, these may represent risks to public health. For example, chlorate and bromate contained in some sodium hypochlorite solutions may cause exceedance of guidelines. Chlorate concentrations in hypochlorite solutions should be monitored regularly, as concentrations increase in storage. The rate of increase is much higher at the elevated temperatures prevalent in tropical countries, and sodium hypochlorite solutions should preferably be stored out of sunlight at low temperature and not for long periods. On-site generation from salt is a desirable option. Bromate is also commonly introduced in the production of sodium hypochlorite from sodium chloride solutions or seawater. The concentration will vary according to the bromide concentrations in the source materials. Hence, similar to conventional water treatment plants, desalination plants must have appropriate facilities to store chemicals in a manner that minimizes the formation of breakdown products. Adequate tankage and temperature control are features that need to be addressed during planning and design. The correct use of these facilities, maintenance of chemical stocks, and control of temperature should be monitored regularly by operators of desalination facilities.

One way of ensuring the purity of delivered chemical feedstocks is to use only chemicals certified for use in the production of drinking water. International certification programs provide confidence in the quality of products. Certification programs provide formulation details, analytical data, including contaminant concentrations, and information on maximum use limits to avoid the introduction of deleterious levels of contaminants into treated water. Programs can also include annual inspection of production facilities as well as inspection and certification of storage and transport systems used to deliver chemicals. This means that storage facilities at production facilities, depots, and transfer stations are checked to make sure that the chemicals can be properly stored and transferred. Depots and transfer stations should be protected from insects and small animals, including birds.

Formal certification of chemicals to international standards will also provide confidence in the purity of products. Chemical suppliers should be evaluated and selected on their ability to supply products in accordance with required specifications. Documented procedures for the control of chemicals, including purchasing, certification, delivery, handling, storage, and maintenance, should be established, and adherence to these procedures should be monitored. Technical specifications should be included in purchase contracts. Contract terms should specify minimum quality requirements as well as certification requirements for each load of chemicals as delivered. Responsibilities for testing and quality assurance of chemicals (supplier, purchaser, or both) should be clearly defined in purchase contracts. If operators are to undertake quality assurance by periodically checking the purity of the chemicals being delivered, access to analytical services will be required either in their own or in a qualified external laboratory.

5.5.2 Testing Equipment, Laboratories, and Methods

Monitoring can be undertaken using online instruments, field kits, and laboratory-based analyses, depending on the application of the data and the precision and accuracy requirements. In all cases, the effectiveness and value of monitoring programs are dependent on accuracy. This requires the application of quality control and assurance procedures. These should include

- Maintenance and regular calibration of online instruments and field kits. Chemicals used in these instruments and kits should be stored under appropriate conditions, and results obtained should be periodically checked by comparison with laboratory-based analyses.
- Assurance of the accuracy and representative nature of water samples. Guidance on sample collection is provided in International Organization for Standardization (ISO) Standard 5667.

- Regular assessment of the competence and accuracy of testing laboratories. General guidance on quality assurance for analytical laboratories is provided in *Water Quality Monitoring: A Practical Guide to the Design and Implementation of Freshwater Quality Studies and Monitoring Programmes* (Bartram and Ballance 1996).

An important issue for desalination facilities is the selection of appropriate testing equipment and testing methods. Equipment and methods used to monitor freshwater sources of drinking water may not be suitable or provide accurate results when used with high-salinity water.

5.6 MONITORING PLANS AND RESULTS

Documentation is an essential component of the WSP, and operational monitoring and verification programs should be documented in a consolidated plan. The plan should include

- Parameters to be monitored
- Sampling locations and frequencies
- Sampling methods and equipment
- Schedules for sampling
- Methods for quality assurance and validation of sampling results
- Requirements for checking and interpreting results
- Responsibilities and necessary qualifications of staff
- Requirements for documentation and management of records, including how monitoring results will be recorded and stored
- Requirements for reporting and communication of results

5.7 SURVEILLANCE

Surveillance of desalinated water supplies should follow the same principles applied to all drinking water supplies. The purpose of surveillance is to assess the safety and acceptability of water supplies, and surveillance preferably should be undertaken by an agency or authority that is independent of the water supplier. In most countries, the agency responsible for surveillance is the ministry responsible for public health or environment. In some cases, the responsibilities could be shared by or delegated to provincial or state agencies or environmental health departments within local government. In other countries, the environmental protection agency may be assigned responsibility. For desalination systems, environmental protection agencies are also likely to regulate performance and undertake surveillance relating to discharges and prevention of environmental impacts.

Surveillance should involve audit-based activities and may also include direct testing. Direct testing requires the surveillance agency to have access to analytical facilities as well as having the capacity to collect samples and interpret results. Where direct testing is performed, it should complement other verification testing and follow the same approach in terms of location and parameters. However, sampling frequencies may be lower than those used in verification undertaken by water suppliers.

Auditing should include examination and assessment of

- The design and implementation of the WSP.
- Records to ensure that system management, including operational monitoring and verification, is being carried out as described in the WSP.
- Results of operational monitoring, including compliance with target criteria and operational limits.
- Results of verification monitoring, including compliance with water quality standards and guideline values.
- Incident responses, including implementation of reporting requirements and corrective actions.
- Supporting programs.
- Strategies for improvement and updating of the WSP.

Auditing can also include sanitary inspections.

5.8 REGULATION

Appropriate legislation, regulations, and standards support the provision of safe and acceptable drinking water. They provide a framework to assist and guide water suppliers in identifying requirements and responsibilities that need to be met in designing, installing, and operating drinking water supplies. They also provide benchmarks by which the activities of water suppliers and the quality of water supplied to consumers can be assessed, as well as penalties for noncompliance if necessary to encourage proper performance.

Legislation should include provisions dealing with

- Management of drinking water supplies
- Drinking water standards and guidelines
- The responsibilities of water suppliers, including notification requirements in the event of incidents and events that may threaten public health
- Surveillance of drinking water supplies, including
    - The identity, functions, and responsibilities of the surveillance agency
    - The authority to undertake surveillance
    - Powers to enforce regulations and standards and to require remedial action in the event of noncompliance
    - The responsibility and authority to issue public advice when drinking water supplies are considered to represent an unacceptable risk to public health
- Protection of sources of drinking water, including protection zones around marine or brackish water intakes to desalination plants or around groundwater bores used as sources for desalination plants.

Other issues that should be included in either legislation or associated codes of practice and technical regulations include the quality and type of materials and chemicals used in the production and supply of drinking water as well as construction and plumbing standards, including provision of backflow prevention and cross-connection control. The annex at the end of this chapter is a sample case study including a number of these regulatory features.

5.9 MONITORING RECOMMENDATIONS

- Monitoring should be conducted in accordance with the principles of WSPs and should include operational monitoring, verification, and surveillance.
- Operational monitoring should be the focus of testing programs and is required to assess the effectiveness of the control measures that are used to ensure that safe drinking water is produced from saline and brackish waters.
- A summary of recommended monitoring requirements for the control measures identified in Chapter 2 is provided in Table 5.1. Indicative frequencies are presented for large (>38,000 m^3/day) and small plants (<38,000 m^3/day).
- The range of parameters and extent of operational monitoring will depend on risk assessment, including
    - The type of desalination process being used
    - The nature, stability, and quality of the source water
    - The size of the desalination system
- Where necessary, operational monitoring requirements should be prioritized on the basis of risk.
- The quality of operational monitoring will depend on the capacity of testing facilities and equipment and application of quality assurance/quality control procedures.
- Verification should follow the same principles as those applied to other types of drinking water systems. However, increased attention may need to be paid to the presence of brominated DBPs, particularly when seawater is the source. In addition, owing to the potentially aggressive nature of desalinated water, verification testing should include corrosion products arising from distribution systems and household plumbing.
- Verification should ensure that safe and acceptable water is delivered to consumers. In addition to testing of water from distribution systems for health-related parameters, verification should include assessment of acceptability. This could include testing for TDS and pH as well as monitoring consumer complaints.
- Surveillance should follow the same principles as those applied to other types of drinking water systems. It can involve audit-based activities and direct testing. Surveillance should preferably be undertaken by an agency or authority that is independent of the water supplier.
- Appropriate legislation, regulations, codes of practice, and technical documents support the provision of safe drinking water and provide a framework to guide water suppliers in identifying requirements and responsibilities.

5.10 RESEARCH NEEDS

- Equipment and methods used for freshwater sources may not be suitable or accurate when used with high-salinity water. Analytical methods tailored for use with seawater and brackish water sources of desalinated supplies need to be developed.
- Improved methods for online operational monitoring of membrane integrity are required.

REFERENCES

Ainsworth, R. 2004. *Safe piped water: Managing microbial water quality in piped distribution systems.* London: IWA Publishing on behalf of the World Health Organization. http://www.who.int/water_sanitation_health/dwq/924156251X/en/index.html.

Bartram, J., and R. Ballance. 1996. *Water quality monitoring: A practical guide to the design and implementation of freshwater quality studies and monitoring programmes.* London: E & FN Spon on behalf of the United Nations Educational, Scientific and Cultural Organization, the World Health Organization and the United Nations Environment Programme. http://www.who.int/water_sanitation_health/resourcesquality/wqmonitor/en/index.html.

Thompson, T., J. Fawell, S. Kunikane et al. 2007. *Chemical safety of drinking-water: Assessing priorities for risk management.* Geneva: World Health Organization. http://whqlibdoc.who.int/publications/2007/9789241546768_eng.pdf.

WHO. 1976. *Surveillance of drinking-water quality.* Geneva: World Health Organization. http://www.who.int/water_sanitation_health/dwq/surveillance/en/index.html.

WHO. 2008. *Guidelines for drinking-water quality.* 3rd ed. incorporating first and second addenda. Geneva: World Health Organization. http://www.who.int/water_sanitation_health/dwq/gdwq3rev/en/index.html.

WHO/WPC. 2006. *Health aspects of plumbing.* Geneva: World Health Organization and World Plumbing Council. http://www.who.int/water_sanitation_health/publications/plumbinghealthasp/en/index.html.

ANNEX: CASE STUDY—REGULATION

The enabling law requires system operators, including water producers, to comply with regulations dealing with supply of safe drinking water together with consideration of plant performance and water conservation. The regulations require

- Licensees to ensure compliance with water quality standards
- That water in major pipelines and trunk mains be not contaminated and be of drinking water quality
- That water supplied to premises be safe and acceptable
- Conservation and efficient use of water
- Reduction of waste and overconsumption
- Publication of water quality information
- That products and processes used be approved for the production of drinking water
- Water to be monitored
- Notification of accidents or abnormal incidents

Three regulations have been established to deliver these outcomes, as described in the following text.

1. WATER QUALITY REGULATIONS

The water quality regulations deal with safety and acceptability, use of products, quality control, provision of information, standards, and sampling frequency. Sampling and testing requirements are based on characteristics of the water system, including source of water, treatment processes, volumes of water produced, storage facilities, and distribution networks.

Sampling requirements are illustrated diagrammatically in Figure 5.1, with parameters divided into six groups (A–F). The parameters included in these groups are shown in Table 5.2.

MSF is the predominant thermal process employed in the region, with less than 15% utilizing MED. Little groundwater is treated using membrane-based processes, and this technology is more likely to be applied to desalination of seawater. In this case, testing for Group D and E parameters may not be required at the plant.

Guidelines are provided on the selection of sampling locations and the collection of samples. For example, it is recommended that treatment facilities incorporate a storage tank with a 24-hour detention time to ensure that mixing and blending are achieved before supply. It is also recommended that water quality samples be collected from this tank.

Examples of sampling frequencies for an MSF plant producing about 1.8 million cubic meters per day are as follows:

- Group A (physical): 360 samples per year
- Group B (chemical): 12 samples per year
- Group C (trace elements): 12 samples per year
- Group F (microbiology): 48 samples per year

In addition to the parameters and standards established to define water quality, there are also requirements relating to minimizing corrosiveness.

Generally, the quality of distillate from MSF processes is as follows:

- TDS: ≤25 mg/L
- pH: 5.5–6.5

The product water must meet the following requirements to ensure that it is noncorrosive in nature:

- TDS: 100–1,000 mg/L
- Chlorine: 0.2–0.5 mg/L at delivery to consumers
- LSI: 0–0.3 (positive)
- pH: 7.0–8.5
- Calcium carbonate: maximum 200 mg/L

In practice, this means that thermal treatment plant operators need to provide post-treatment blending or remineralization. This can include blending with filtered seawater or the addition of approved chemicals. Distillate is slightly acidic, and the pH needs

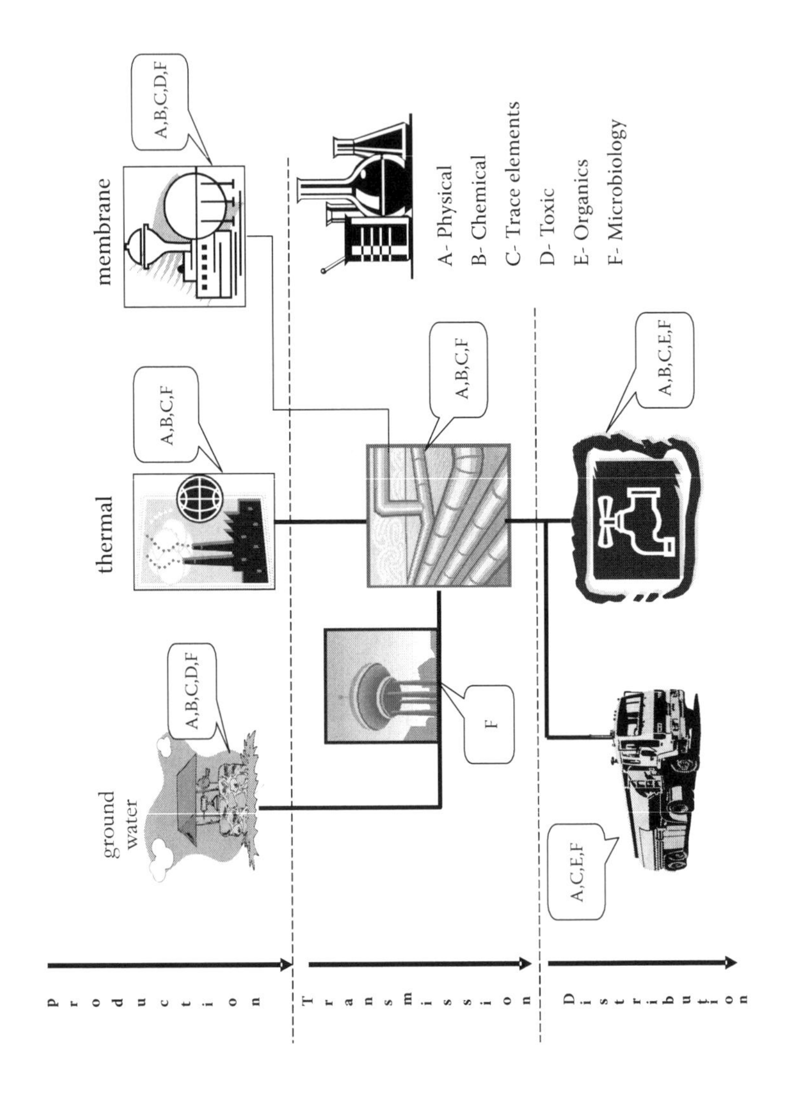

FIGURE 5.1 Water quality sampling guide.

TABLE 5.2
Groups of Water Quality Parameters

Group	Classification	Parameters
A	Physical	Color, turbidity, odor, taste, TDS, calcium hardness, total hardness, chlorine, conductivity, pH
B	Chemical	Sulfate, magnesium, sodium, potassium, chlorides, nitrate, nitrite, ammonium, TOC, dissolved or emulsified hydrocarbons, mineral oil, aluminum, iron, manganese, copper, zinc, phosphorus, fluoride
C	Trace elements	Arsenic, cadmium, cyanides, chromium, mercury, nickel, lead, antimony, selenium, barium, boron
D	Toxicants	Endrin, lindane, methoxychlor, 2,4-dichlorophenoxyacetic acid, 2,4,5-trichlorophenoxypropionic acid, phenols, heptachlor, aldrin, dichlorodiphenyltrichloroethane, chlordane, dieldrin, heptachlor epoxide
E	Organics	Trichloroethene, tetrachloromethane, tetrachloroethene, chloroform, polycyclic aromatic hydrocarbons, 1,2-dichloroethane, benzene, benzo(*a*)pyrene, bromoform, dichloromethane, bromodichloromethane, chlorobenzene
F	Microbial	Total coliforms, *E. coli* or thermotolerent coliforms, enterococci, HPC

to be raised by adding lime or sodium hydroxide. Corrosion inhibitors may also be added. Finally, the distillate requires disinfection to maintain microbiological quality.

With membrane desalination, the required posttreatment may be limited to the adjustment of pH and disinfection. However, if the LSI indicates that the water is corrosive, calcium hardness and alkalinity may need to be adjusted.

The regulations incorporate a number of criteria relating to quality control, including

- Maintenance of plants and prevention of contamination during and after completing repairs
- Use of process chemicals and inspection procedures
- Certification of laboratory equipment and testing methodology
- Additional water quality testing before supply following restarting of thermal desalination units after long stoppages or major overhauls for products of corrosion, including iron, copper, nickel, cadmium, and other trace elements

2. DRINKING WATER SUPPLY REGULATIONS

The water supply regulations address issues associated with customer storages and plumbing. The regulations include criteria relating to construction materials, tank fittings, tank cleaning, disinfection, and inspection provisions. The regulations also require fitting of water meters and automatic control devices to prevent tank flooding and back-contamination.

3. INCIDENT REPORTING AND INVESTIGATION REGULATIONS

System operators are required to maintain contingency plans for bacteriological contamination and hydrocarbon pollution and report any such incidents to the regulatory authority.

6 Environmental Impact Assessment of Desalination Projects

Sabine Lattemann, Khalil H. Mancy, Bradley S. Damitz, Hosny K. Khordagui, and Greg Leslie

CONTENTS

The United Nations Environment Programme's Regional Office for West Asia and the World Health Organization's (WHO) Regional Office for the Eastern Mediterranean published an expanded version of this chapter in 2008 (UNEP 2008). The UNEP (2008) document is divided into three parts. In part A, an introduction to the concept, methodology, and practice of environmental impact assessment (EIA) for desalination projects is given, and a 10-step EIA approach is proposed. Part B outlines a possible modular structure of an EIA report and gives an overview on a wide range of thematic issues that may be relevant to desalination projects. Part C discusses the potential impacts of desalination plants on the environment, based on a comprehensive literature review, and evaluates the identified impacts in terms of their significance and relevance for EIA studies. In the following, the main results from part A are summarized.

An EIA is a procedure that identifies, describes, evaluates, and develops means of mitigating potential impacts of proposed activities on the environment. The main objective of an EIA is to promote environmentally sound and sustainable development through the identification of appropriate mitigation measures and alternatives. In an EIA, information on the environmental consequences of a project is provided to the public and to decision makers. With the help of the EIA, a decision can be reached that balances the societal and environmental impacts and benefits. A detailed EIA is often required for large infrastructure projects, such as a water treatment plant or a dam serving one or more large urban centers. For smaller projects, a simplified EIA may be warranted as a result of the limited potential of the project to cause significant environmental impacts. In principle, EIAs for desalination projects should not differ in terms of complexity and level of detail from those for other infrastructure projects and water supply systems. Depending on the proposed project, it is incumbent on national authorities to individually define the need, scope, and complexity requirements for each EIA study.

Project EIAs are to be distinguished from strategic environmental assessments—or strategic (program) EIAs—that are undertaken for strategic plans, policies, or management programs. The main objective of a strategic EIA is to ensure that potential impacts are addressed during preparation and before adoption of a new plan, policy, or program. Strategic EIAs will not make EIAs at the project level dispensable; rather, they complement each other. Although this chapter will not further address strategic EIAs, it is emphasized that these could be adequate instruments to manage water supply planning on a regional or national scale. The most relevant plans to address desalination projects along with other water supply alternatives are integrated water resources management and integrated coastal zone management plans.

EIAs are not limited to environmental aspects, but typically address all potential key positive and negative impacts of new projects, plans, or activities on "humans and the environment." They consequently relate to many facets of science. In general, an EIA for a water supply project, including that for a desalination project, should predict the impacts related directly or indirectly to the implementation of the project. This may require an interdisciplinary approach covering relevant issues of marine and terrestrial ecology, hydrology, geology, and other disciplines. Taken a step further in relating potential impacts on people and communities, it may also be necessary to consider human health and socioeconomics. Where appropriate, these should take into account gender- and age-specific effects and variations among the potentially affected population or community, such as social or ethnic affiliations of subgroups. Public participation is also considered a fundamental element of EIAs in order to involve the public in the evaluation of potential impacts and in decision making.

The environmental impact of desalination and its discharges should also be assessed in the context of the environmental impacts of water supply alternatives that may be used instead of desalination. Desalination projects are typically driven by the limited availability of alternative lower-cost water supply resources such as groundwater and recycled or fresh surface water (rivers, lakes, etc.). However, environmental impacts may also result from continuation of those water supply practices. For example, overpumping of coastal aquifers over the years has resulted in a significant increase in the salinity of the groundwater and damaged freshwater aquifers. In some arid areas, transfers of fresh water from a traditional water supply source, such as a river, river delta, or lake, have impacted the eco-balance in this water source to such an extent that the long-term continuation of such a water supply practice may result in significant and irreversible damage to the ecosystem of the traditional freshwater supply source. In such cases, the environmental impacts of the construction and operation of a new seawater or brackish water desalination project should be weighed against the environmentally damaging consequences from the continuation or expansion of the existing freshwater supply practices. Waste streams generated from desalination plants, with the exception of the high-salinity reject water, are similar to the waste streams generated by conventional water treatment plants and water reuse facilities. Water reclamation plants also generate waste streams that usually contain the same chemicals used for desalination and may have an elevated content of anthropogenic waste substances that may have impacts on the marine environment. Wastewater reuse for beneficial purposes is another factor to be considered among the options and combinations of options that exist.

With the context so broad, the present chapter cannot fully encapsulate the entire spectrum and depth of implications of all desalination projects on a worldwide scale. Rather, it provides a general overview of the potential impacts of a desalination project on the environment. Emphasis is placed on the impacts that are specific to desalination projects, such as the impacts of reject streams and chemical additives on the marine environment, whereas impacts common to other infrastructure projects are not discussed in detail.

This chapter further seeks to provide guidance on how to carry out an EIA (see Section 6.2). The proposed methodology is not limited to desalination plants, but can be applied to other water infrastructure projects in a similar manner. Based on the information provided, which issues are relevant for an individual desalination project should be decided on a case-by-case basis, and an individual approach should be developed to carry out the EIA.

In addition to the obvious benefits of a supply of high-quality water, desalination, similar to other water supply alternatives, will consume considerable community resources. This includes economic and social capital, access to coastal land, energy consumption, and ongoing revenue to operate and maintain the facility. Desalinated water should thus be regarded as a community asset and a valuable resource. In addition to considering the measures outlined in this chapter and in the UNEP (2008) guidance to assess and mitigate the potential impacts of the production process, it is also incumbent upon communities to value the water produced by desalination by nonwasteful use and by looking for opportunities of multiple use before the water is discharged.

Finally, although there is at least a 50-year history of large-scale desalination projects, the present knowledge of the environmental, socioeconomic, and human health implications of desalination and its alternatives (water reuse, extended use of limited traditional water supply sources, and conservation) could be expanded. This, and the fact that EIAs are undertaken before projects are implemented, means that although they are based on detailed analyses, EIAs can only give a prognosis of the expected impacts based on the information available. It is therefore important to clearly identify any knowledge gaps in the EIA and to apply the precautionary principle as defined by the United Nations (UN) in the evaluation of potential impacts. Further research should be initiated, and monitoring data from existing facilities and EIA results should be made available to a wider public to improve the understanding of the actual impacts of all water supply alternatives, including desalination, reuse, conservation, and the use of traditional water supply sources.

6.1 POTENTIAL ENVIRONMENTAL IMPACTS OF DESALINATION PROJECTS

Chapter 2 of this book includes a discussion of the potential impacts of desalination projects on the environment and references for further reading. It describes impacts on the marine environment that are specific to desalination projects, in particular the impacts of reject streams and chemical additives. Impacts common to many development projects, such as surface sealing or air emissions, are not covered. It is assumed that common effects are sufficiently known and that information is readily available from other sources, such as relevant literature or other development projects. There are still many knowledge gaps and uncertainties regarding the actual impacts of desalination projects, as monitoring results of operating plants are available to only a limited extent. Also, a wide variety of project- and site-specific impacts may occur. The list of potential environmental concerns discussed in the UNEP (2008) guidance can thus not be complete or final, and not every described effect will apply to each individual project. Further research is required, and the provision of monitoring results to a wider audience is highly recommended.

6.2 CONCEPT AND METHODOLOGY OF EIA IN GENERAL AND IN DESALINATION PROJECTS

6.2.1 Introduction to Development of an EIA

EIA is a systematic process used to identify, evaluate, and develop means of mitigating potential effects of a proposed project prior to major decisions and commitments being made. This process applies for all wastewater, water reuse, and water treatment projects, including desalination plants. It usually adopts a broad definition of "environment," considering socioeconomic as well as public and environmental health effects as an integral part of the process. The main objectives of EIAs are to provide information on the environmental consequences for decision making and to promote environmentally sound and sustainable development through the identification of appropriate alternatives and mitigation measures (UNEP 2002).

The three central pillars of an EIA are

1. The establishment of environmental, socioeconomic, and public health baseline data for the project site before construction. A prognosis of the "zero alternative" is given, which is the expected development of the project site without project realization.
2. The prediction and evaluation of potential—direct and indirect—environmental, socioeconomic, and public health impacts of the proposed project.
3. The identification of appropriate alternatives and mitigation measures to avoid, minimize, remediate, or compensate for any environmental, socioeconomic, and public health impacts resulting directly or indirectly from the project.

In essence, EIA of desalination projects is a systematic process that examines the environmental, socioeconomic, and health effects during all life cycle stages of the project (i.e., during construction, commissioning, operation, maintenance, and decommissioning of the plant). The impacts of a proposed desalination project are those alterations in environmental conditions compared with what would have happened had the project not been undertaken (zero alternative).

6.2.2 Systematic EIA Process for Desalination Projects

The EIA process is generally marked by three main phases (Figures 6.1 and 6.2):

1. The initial, or pre-EIA phase, includes screening and scoping of the project.
2. The main EIA phase refers to the actual EIA, including the establishment of baseline data, the prediction and evaluation of potential impacts and the identification of appropriate alternatives and mitigation measures (see "three pillars" earlier).
3. The final EIA phase involves decision making and a review of the EIA process.

In the following, a 10-step process is proposed for conducting EIAs for desalination projects. It should be noted that deviations from the outlined process may occur in practice. Single steps may not always be clearly limitable; some steps may overlap or may be interchanged. The EIA procedure should thus be understood as a continuous and flexible process.

6.2.2.1 Step 1: Screening of the Project

Screening is the process by which a decision is taken on whether or not an EIA is required for a particular project. It will ensure that a full EIA is performed only for projects with potentially significant adverse impacts or where impacts are not sufficiently known. Screening thus involves making a preliminary determination of the expected impact of a proposed project on the environment and of its relative significance. A certain level of basic information about the proposal and its location

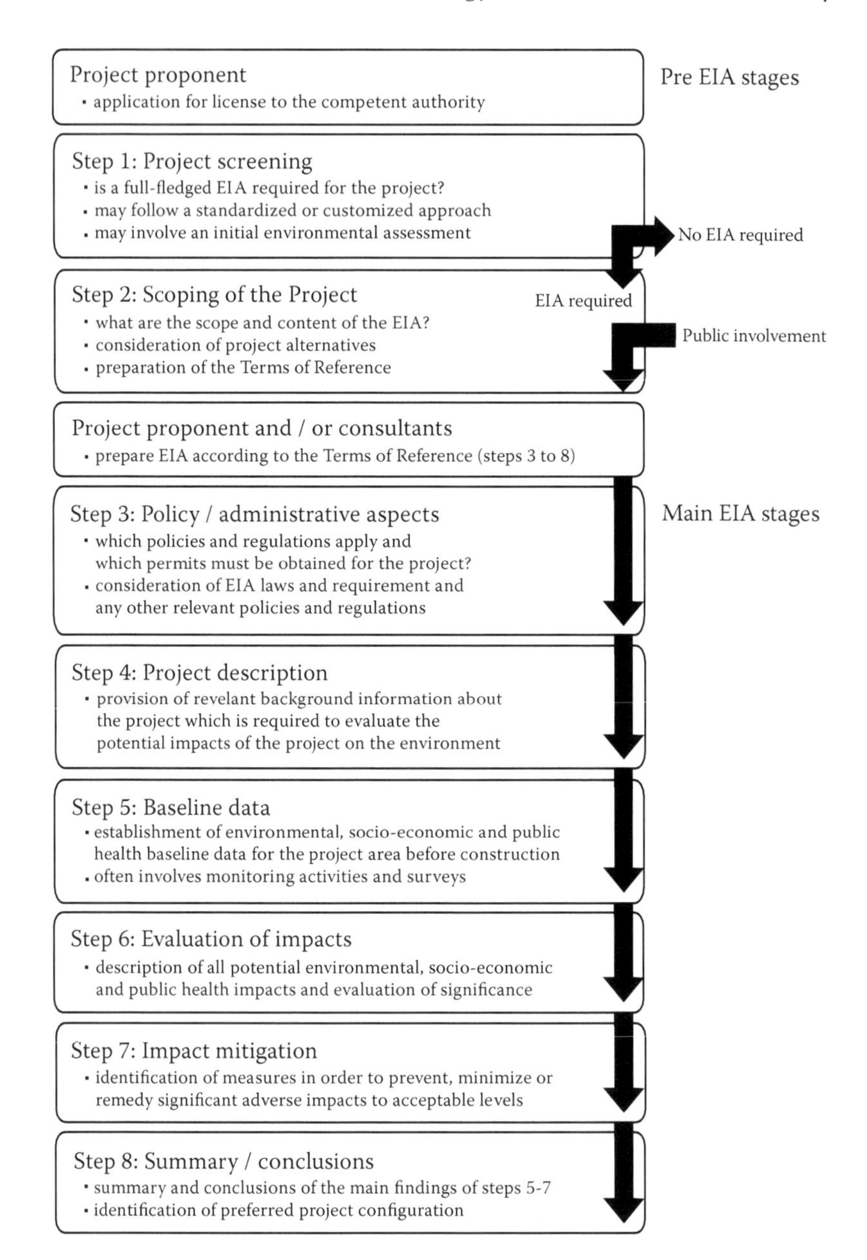

FIGURE 6.1 Ten-step EIA process: Scoping, screening, and main EIA phase.

is required for this purpose. The screening procedures can be broadly classified into two approaches: a *standardized* approach, in which projects are subject to or exempt from EIA as defined by legislation and regulations; and a *customized* approach, in which projects are screened on a case-by-case basis, using indicative guidance (UNEP 2002).

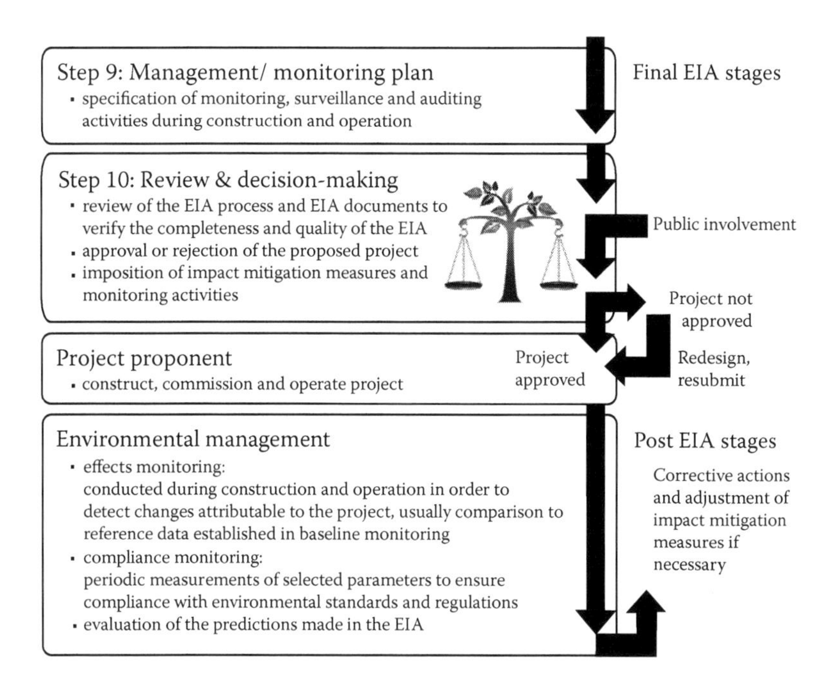

FIGURE 6.2 Ten-step EIA process (continued): EIA decision phase and follow-up activities.

6.2.2.1.1 *Standardized Approach*

Many states have implemented EIA laws and procedures that facilitate the screening process by defining for which project categories an EIA is required, such as the following:

- "Mandatory" or "positive" lists that include projects always requiring EIA (e.g., major projects, possibly large cogeneration plants for electricity and water).
- Project lists that define thresholds and criteria above which EIA is required (e.g., a desalination plant with more than 20,000 m^3/day of production capacity).
- "Exclusion" or "negative" lists that specify thresholds and criteria below which EIA is never required or below which a simplified EIA procedure applies (e.g., a desalination unit with less than 4,000 m^3/day of production capacity).

A class screening may be undertaken for small-scale projects that are routine and replicable, if there is a reasonably sound knowledge of the environmental effects and mitigation measures are well established. For example, class screening could be applicable to small-scale, stand-alone desalination systems such as those for hotels, small residential and recreational communities, industrial facilities, military installations, etc. The regulations for water treatment facilities in general and for desalination plants in particular may vary considerably in different states. If the categorization of projects in general or of desalination plants in particular has not been undertaken or if a proposed desalination project is on the borderline of a threshold, the project would need to be screened on a case-by-case basis (customized approach).

6.2.2.1.2 Customized Approach

Individual screening does not necessarily require additional studies, but can be conducted on the basis of indicative guidance—for example, indicators and checklists. These are intended to be used quickly by people with the qualifications and experience typically found in competent authorities or environmental consultant companies, using the information that is readily available about the project and its environment.

The World Bank (1991) categorization of projects may allow a first, broad screening of desalination plants based on a few common indicators, such as the type, size, and location of the project, environmental sensitivity, and likely health and social effects on the local population:

- *Category A—full EIA required*: Projects likely to have significant adverse environmental impacts that are significant (i.e., probably irreversible, affect vulnerable ethnic minorities, involve involuntary resettlement, or affect cultural heritage sites), diverse, or unprecedented or that affect an area broader than the sites of facilities subject to physical works (e.g., dams and reservoirs, large-scale industrial plants, port development, thermal power and hydropower development).
- *Category B—limited EIA to identify and incorporate suitable mitigation measures*: Projects likely to have adverse environmental impacts that are less significant than those of category A, meaning that few, if any, of the impacts are likely to be irreversible, that they are site specific and that mitigation measures can be designed more readily than for category A projects (e.g., small-scale aquaculture, renewable energy, rural electrification, rural water supply, and sanitation).
- *Category C—no EIA*: Projects that are likely to have minimal or no adverse environmental impacts.

Comprehensive indicator lists or checklists can be used for screening, such as the checklists prepared by the European Union (EU 2001a) as part of the EIA directive framework (European Commission 1997). They include a list of questions referring to the project and its environment, which will help to answer the question: is the project *likely* to have a *significant effect* on the environment? There is no specific rule that can be used to decide whether the results of screening should lead to a positive or negative screening decision (i.e., that an EIA is or is not required). As a general principle, the greater the number of positive answers and the greater the significance of the effects identified, the more likely it is that an EIA is required. Uncertainty about the occurrence or significance of effects should also point toward an EIA, as the EIA process will help to clarify the uncertainty. If the need for EIA has been affirmed, scoping follows as the next step.

6.2.2.1.3 Preliminary EIA Study

In some EIA systems, screening is considered a flexible process that can be extended into a preliminary form of an EIA study (often termed *preliminary* or *initial environmental assessment*). This is typically carried out in cases where the environmental impacts of a proposal are uncertain or unknown—for example, new technologies or undeveloped areas (UNEP 2002). If a preliminary assessment is undertaken to assist in the screening decision, the information from the preliminary assessment can also be used for scoping and later in the actual EIA process.

6.2.2.1.4 Documentation of Screening Results

After a formal decision has been made by the competent authority, an official screening document is typically prepared that records the screening decision and provides an explanatory statement for this decision. It may be extended into a short screening report that also gives the results of the preliminary assessment and can be used to prepare the scoping document for public dissemination in the following stage.

6.2.2.2 Step 2: Scoping of the Proposed Desalination Project

Scoping is the process of determining the content and extent of the EIA. The Terms of Reference (ToR), which are elaborated in the process, provide clear instructions to the project proponent on the information that needs to be submitted to the competent authority for EIA and the studies to be undertaken to compile that information.

Scoping is a crucial step in EIA because it identifies the issues of importance and eliminates those of little concern. In this way, it ensures that EIAs are focused on the significant effects and do not involve unnecessary investigations that waste time and resources. The process is completed with preparation of the ToR. However, experience shows that the ToR should be flexible to some degree, as they may need alteration as further information becomes available and as new issues emerge or others are reduced in importance (UNEP 2002). The range of issues in the following sections should be considered when planning the scoping phase and preparing the ToR.

6.2.2.2.1 Consideration of Alternatives

The consideration of alternatives to a proposal, such as alternative technologies or sites, is a requirement of many EIA systems. It should be understood as a dynamic process that starts early in project planning and continues throughout the EIA process and decision making.

Alternatives to a proposal can be generated or refined most effectively in the early stages of project development. At the stage of scoping, a number of alternatives are typically identified, which are subsequently evaluated in the EIA. Alternatives may also be identified later on in the process, especially at the stage when impact mitigation measures are elaborated. The EIA should be open to new emerging alternatives at all times, whereas previously considered options are possibly abandoned in the light of new information. The aim is to identify the best practicable option under environmental, socioeconomic, and human health criteria that is also technically and economically feasible. It is important that the consideration of alternatives during an EIA is not reduced to a superficial and meaningless exercise.

6.2.2.2.2 Selection of the Project Site

Environmental, socioeconomic, and public health impacts resulting from the construction and operation of a desalination plant are in large part dictated by the location of the facility and its associated infrastructure. Therefore, proper site selection for a desalination plant during the planning process is essential for impact minimization.

Site selection typically takes place in the early stages of a desalination project and leads to the identification of a preferred site and possibly one or two alternatives. An EIA, usually accompanied by a site-specific monitoring program, will then be carried out for the identified location. In many cases, the competent authority will attach conditions to project approval, such as the requirement to implement certain mitigation measures in order to minimize impacts on the project site. In some cases, however, the EIA may also come to the final conclusion that the chosen site is not suitable, even if impact mitigation measures are implemented.

To reduce the likelihood of this outcome, site selection should play an important role in project planning. Site selection can take place during a "preliminary" EIA study as part of the screening process (cf. Step 1 above) or during scoping when the EIA requirements are determined. To facilitate the site selection process for desalination plants, public authorities may designate suitable areas in regional development plans or may provide criteria that can be used by project developers. Site selection must be carried out on a case-by-case basis, as there are a large number of site-specific considerations that vary according to the specific operational aspects of each plant. Generally, it is important to consider the following site features:

- *Geological conditions*: Sites should provide stable geological conditions, and there should be little risk that construction and operation of the plant will affect soil and sediment stability.
- *Biological resources*: Ecosystems or habitats should be avoided, where possible, if they are unique within a region (e.g., reefs on an otherwise sandy shoreline) or worth protecting on a global scale (e.g., coral reefs, mangroves); if they are inhabited by protected, endangered, or rare species (even if temporarily); if they are important in terms of their productivity or biodiversity; if they have important biological functions, such as feeding grounds or reproductive areas for a larger number of species or certain key species within a region; or
- If they are important for human food production (fishing or marine harvesting).
- *Oceanographic conditions*: The site should provide sufficient capacity to dilute and disperse the salt concentrate and to dilute, disperse, and degrade any residual chemicals. The load and transport capacity of a site will depend primarily on water circulation and exchange rate as a function of currents, tides and/or surf, water depth, and bottom and shoreline morphology. In general, exposed rocky or sandy shorelines with strong currents and surf may be preferred over shallow, sheltered sites with little water exchange. Water exchange and sediment mobility will affect the residence time of pollutants within the ecosystem and the time of exposure for marine life to these pollutants.

- *Concentrate management*: Environmental impacts associated with concentrate discharge have historically been considered the greatest single ecological impediment in selecting the site for a desalination facility. However, aquatic life impingement and entrainment by the desalination plant intake are more difficult to identify and quantify, and may also result in measurable environmental impacts. It is therefore recommended to conduct site-specific entrainment and impingement studies to investigate the magnitude of potential impacts and to develop and adopt appropriate impact mitigation measures (see Chapter 2 for more details).
- *Collocation with power plants*: Collocation with power plants is recommended not only to minimize entrainment and impingement but also to minimize construction impacts in the coastal zone and to allow for dispersion of the brine. These aspects should also be considered in the site selection process. Also, subsurface intakes are an environmentally friendly option, but may not be feasible in all locations. For this reason, the recommendation should not be narrowed down to a single intake type as the "Holy Grail" (see Chapter 2 for more details).
- *Raw water quality and proximity*: The intake location should ideally provide good and reliable water quality, taking seasonal changes into account, with minimum danger of pollution or contamination, in order to avoid performance problems of the plant or impacts on product water quality. The plant site should ideally be close to the source water intake to minimize land use for pipelines and to avoid passage of pipes through agricultural land, settlements, etc. However, this cannot be generalized, and in some cases it may be more appropriate to locate the plant farther inland, for example, when construction on the shore is not possible for certain reasons (e.g., use of beaches, nature reserves, geological instability, etc.).
- *Proximity to water distribution infrastructure and consumers*: The site should ideally be close to existing distribution networks and consumers to avoid construction and land use of pipelines and pumping efforts for water distribution. However, impairment of nearby communities (i.e., consumers) by visual effects, noise, air pollution, or other environmental health concerns should be avoided.
- *Vicinity of supporting infrastructure*: The site should allow easy connection with other infrastructure, such as power grid, road and communication network, or may even allow the co-use of existing infrastructure, such as seawater intakes or outfalls.
- *Conflicts with other uses and activities*: The site should ideally provide no conflict or as little as possible with other existing or planned uses and activities, especially recreational and commercial uses, shipping, or nature conservation.

6.2.2.2.3 Public Involvement

Public participation is a mandatory requirement in the planning and implementation of development projects and an inherent component of the EIA process, especially for scoping. As a general rule, the public should be involved as early as possible and continuously throughout the EIA process. The overall goal is the involvement of the public in decision making. This is based on fundamental premises of democratic societies, such as transparency of decision making and equity among the affected populations in terms of ethnic background and socioeconomic status.

Public involvement in desalination projects is aimed at

- Informing the public about the desalination project.
- Gathering a wide range of perceptions of the proposed desalination project, addressing the benefits, potential public health, environmental and socio-economic impacts, and their short- and long-term implications.
- Presenting a discussion of project alternatives, including water conservation and recycling.
- Providing information on the value of the desalinated water and the extent of the likely community investment.
- Taking advantage of the knowledge of indigenous and local communities about their living environment.

The overall benefits of public involvement are to

- Develop a partnership with the community, which is critical for project sustainability.
- Address and, where applicable, dispel subjective public doubts and concerns about the project.
- Develop trust and working relationships among the stakeholders, including the affected communities, particularly vulnerable groups, developers, planners, local and national governments, decision makers, nongovernmental organizations (NGOs), and networks of people and organizations.
- Ensure that important issues are not overlooked when the ToR are prepared, thus providing for the comprehensiveness, integrity, quality, and effectiveness of EIA.

6.2.2.2.4 Human Health Impact Assessment

EIAs, as widely required by national legislation and international agencies, offer integrated analyses of potential impacts of development projects on all components of the environment, including human health. There has been recent emphasis on the necessity to delineate the health effects of environmental impacts—as stated in the European Directives (European Commission 1997, 2001) and the Espoo Convention on EIA (UNECE 1991)—on directly or indirectly affected populations. When conducting scoping for a desalination project, relevant human health effects should therefore be identified.

The human health component should be broadly addressed in EIAs for desalination projects, relying on readily available information. This includes community health determinants, such as incidences of disease, public information and concerns, and traditional knowledge of the local inhabitants and indigenous population. Baseline information on health and quality of life needs to be established in order to assess the significance of the potential effects of environmental impacts. Potential environmental health impacts should be prioritized, with corresponding indicators and risk factors. Both positive and negative health effects should be delineated, for the public at large as well as for vulnerable groups. Where there are specific concerns with exposure to certain toxic emissions or infectious agents, the scientific literature should be searched for relevant published studies and epidemiological investigations. This is usually sufficient to address concerns with the potential health impact. Most EIA assessments rely on existing information. Except for large projects, it is often too expensive and too time consuming to generate new health information within the time frame allotted to develop the EIA.

If the mineral composition (e.g., calcium, magnesium, sodium, fluoride, total dissolved solids) of the desalinated water will be significantly changed from that of the current drinking water, it would be appropriate to conduct before-and-after studies of certain endpoints to determine any potential positive or negative consequences. Desalination has been used in some parts of the world for many decades, and this experience potentially provides a basis for total diet and water epidemiological studies of various health outcomes, including cardiovascular disease (CVD), osteoporosis, metabolic syndrome, and incidence of dental caries in the population. Such studies, if properly controlled and designed with proper consideration of potential confounding factors, would be of considerable value in ensuring the safety of desalinated water for long-term consumption and appropriate stabilization techniques. WHO is recommending that before-and-after studies of acute CVD mortality be conducted in drinking water supplies that are undergoing changes in calcium and magnesium content (Cotruvo 2009; see Chapter 3 for additional information).

Procedures for environmental health impact assessment within EIAs are described in greater detail in the publication *Environmental Health Impact Assessment of Development Projects: A Practical Guide for the WHO Eastern Mediterranean Region* (Hassan et al. 2005).

6.2.2.2.5 Gender Impact Assessment

Gender mainstreaming is a globally accepted strategy for promoting gender equality. The UN Economic and Social Council (United Nations 1997) defined gender mainstreaming as the process of "assessing the implications for women and men of any planned action, including legislation, policies or programmes, in all areas and at all levels," so that "women and men benefit equally and inequality is not perpetuated." Gender impact assessment is increasingly recognized as an adequate tool for implementing gender mainstreaming. It is usually applied to policies and programs and means to compare and assess, according to gender-relevant criteria, the current situation and trend with the expected development (European Commission 1998). In the same manner that policies and programs may have differential impacts on women and men, many development projects will not be gender neutral. Gender-specific effects may not be easily recognized, but an effort should be made to identify any significant effects that may perpetuate gender inequality.

Water projects, and thus desalination projects, have a high potential for gender-specific effects. "Women play a central part in the provision, management, and safeguarding of water," which is one of four recognized principles of the Dublin Statement on Water and Sustainable Development (ICEW 1992). The consideration and integration of gender-specific effects in EIAs for desalination plants, from scoping to decision making, are thus highly recommended to evaluate the advantages and disadvantages of desalination activity on both sexes. Where appropriate, a distinction in the EIA process should be made between impacts on men and impacts on women.

6.2.2.2.6 *Scoping Procedure*

Scoping procedures may vary considerably in different states. For example, scoping may be carried out under a legal requirement or as good practice in EIA, or it may be undertaken by the competent authority or by the project proponent (EU 2001b). It is recommended that the competent authority take responsibility at least for monitoring of the process, for preparing the minutes and official transcripts of the scoping meetings, for keeping the records of the scoping outcome, and for preparing the ToR. As a generalized approach, the scoping procedure may follow the following steps:

- Based on the information collected during screening, a scoping document containing a preliminary environmental analysis will be prepared. This will specify details on the proposed location of the project, review alternatives, briefly and concisely describe the environmental characteristics of the selected site, and raise potentially significant project-related issues. The scoping document serves as a background document for hearings and discussions.
- The date and venue for the scoping meeting will be set and a provisional agenda prepared. Invitations for the meeting and the scoping document will be sent to collaborating agencies, stakeholder groups, NGOs, experts, and advisers. The scoping meeting will also be announced in public and the scoping document put on display for public inspection. A handout may be circulated, notices posted in communities, and media advertisements arranged to enhance public participation. If the number of potentially interested people and organizations is large, questionnaires requesting written comments should be considered.
- During scoping consultations, a complete list of all issues and concerns should be compiled. These items may then be evaluated in terms of their relative importance and significance to prepare a shorter list of key issues, which can be classified into different impact categories to be studied in the EIA.
- The ToR for EIA will be prepared, including information requirements, study guidelines, methodology, and protocols for revising work.

6.2.2.2.7 *Scoping Tools and Instruments*

When a competent authority or a developer undertakes scoping (EU 2001b), three key questions should be answered:

- What effects could this project have on the environment?
- Which of these effects are likely to be significant and therefore need particular attention in the environmental studies?
- Which alternatives and mitigating measures ought to be considered?

Basic instruments such as checklists and matrices are often used to provide a systematic approach to the analysis of potential interactions between the project and the environment. For example, checklists for scoping have been elaborated by the European Union as supporting information to the European EIA directive framework (European Commission 1997). The scoping checklists provide a list of project characteristics that could give rise to significant effects and a list of environmental characteristics that could be susceptible to significant adverse effects.

6.2.2.2.8 Standardized Scoping Procedure

An effective way of dealing with an increasing number of desalination projects may be to elaborate a standardized scoping procedure and ToR. The scoping process will often involve the same representatives of government agencies, NGOs, consultants, etc. A guideline, elaborated in a collaborative effort between these groups, may routinize the scoping procedure and may establish standards for the environmental studies to be undertaken and the information to be submitted in EIAs for desalination plants. The guideline could thus serve as a blueprint for scoping, which should still allow for project-specific adjustments.

6.2.2.3 Step 3: Identification and Description of Policy and Administrative Aspects

EIAs usually take place within the distinctive legislative frameworks established by individual countries and/or international agencies. It is therefore recommended to gain a deeper insight and understanding of any national policies or international agreements that apply in a country or region and that relate to EIA procedures in general (UNEP 2002).

Moreover, any other policy relevant to the desalination project needs to be identified. Major thematic areas that should be considered when searching the national or international legal system include conservation of nature and biological diversity, control and prevention of pollution, water resources management, and land use and regional planning.

In many jurisdictions, more than one permit will be required to realize a desalination project. The main approval process, which authorizes construction and operation of a plant, will not necessarily replace other existing statutory provisions and permits. For example, the construction and operation of a desalination plant can present a number of safety hazards to plant workers, so a specific workplace safety permit will probably be required and/or a plan must be developed to ensure the occupational safety and health of the workers.

It is important to clarify early in project planning which permits must be obtained and to contact the competent authorities. The permitting process may be facilitated by nominating a "lead" agency, which coordinates the process by involving other agencies and by informing the project proponent about permitting requirements.

6.2.2.4 Step 4: Investigation and Description of the Proposed Desalination Project

A technical project description should be prepared and included in the EIA report. It should form the basis of the EIA process by providing background information on the project that is required to investigate and analyze all potential impacts. The project description should cover the different life cycle stages of construction, commissioning, operation, maintenance, and decommissioning of the desalination plant. It should be succinct, containing all the information necessary for impact assessment but omitting irrelevant or distracting details.

6.2.2.5 Step 5: Investigation and Evaluation of Environmental Baseline

This step will entail assembling, evaluating, and presenting baseline data of the relevant environmental, socioeconomic, and public health characteristics of the project area before construction and including any other existing levels of degradation or pollution.

A reference area with similar baseline characteristics may be selected and surveyed in parallel with which the project site can be compared during project monitoring in order to detect any changes caused by construction and operation activities.

The scope of the baseline studies to be undertaken in an EIA for a desalination project should have been determined during scoping (Step 2) and should be briefly outlined in the EIA report. The baseline studies will often have the following information requirements:

- *Socioeconomic and sociocultural environment*: Population, land use, planned development activities, status of existing water resource management programs (conservation and reuse), community structure, employment, distribution of income, goods and services, recreation, public health, cultural properties, tribal and indigenous people, customs, attitudes, perceptions, aspirations, etc.
- *Public health environment*: Health indices of the populations at risk of being affected by the project (e.g., rates of morbidity, mortality, injuries, accidents, and life expectancy), as well as relevant socioeconomic indicators of the quality of life. It should be noted here that the WHO Constitution (WHO 1948) defines health as the "state of complete physical, mental, and social well-being and not merely the absence of disease or infirmity."
- *Abiotic environment*: Geology, topography, climate, meteorology, ambient air quality, surface water and groundwater quality and hydrology, coastal and marine environmental quality, existing sources of emissions to air, soils, and water, capacity of environmental systems to take up, degrade, dilute, and disperse emissions, noise levels, etc.
- *Biotic environment*: Flora and fauna, including rare and endangered species, sensitive habitats, species of commercial value, species with potential to become nuisances, etc.

6.2.2.6 Step 6: Investigation and Evaluation of Potential Impacts of the Project

In this step of the EIA, a prognosis, description, and evaluation of the potential environmental, socioeconomic, and health impacts of the proposed project are elaborated. Potential impacts can be identified and evaluated by comparing the observed changes in the project area with baseline data from preconstruction or with reference data from a reference site. Reference data from a site with similar environmental characteristics may be particularly useful to identify natural variations or detect other anthropogenic changes that are not attributed to the desalination project.

The magnitude, spatial and temporal range of all identified impacts, and their relative significance should be evaluated in this step. Where possible, an attempt should be made to further distinguish between direct and indirect impacts, immediate and long-term impacts, reversible and irreversible impacts, avoidable and unavoidable impacts, and positive and negative impacts. It is recommended that identified positive and negative effects also be balanced in terms of their societal and environmental costs and benefits. If possible, potential cumulative, transboundary, and growth-inducing effects should be identified and investigated. Careful deliberation about the accuracy of all predictions made in the EIA is recommended. These can only be as accurate and valid as the data and information available. It is therefore necessary to identify any information gaps and deficiencies in the EIA and to assess any uncertainties associated with the prognosis of impacts.

6.2.2.6.1 Application of the Precautionary Principle

The precautionary principle as defined by the UN should be applied in the EIA where uncertainty about impacts exists. The precautionary principle was defined and adopted by the Rio Conference on the Environment in 1992 (UNEP 1992) as Principle No. 15, which states that "in order to protect the environment, the precautionary approach shall be widely applied by States according to their capability. Where there are threats of serious or irreversible damage, lack of full scientific certainty shall not be used as a reason for postponing cost-effective measures to prevent environmental degradation." The precautionary principle therefore requires action to be taken to prevent serious adverse impacts on human health or the environment, even if there is not incontrovertible proof, but where there is strongly suggestive evidence. This does not mean action on the basis of speculation and complete lack of evidence. There is, of course, a balance to be struck between human welfare and environmental protection that will allow sustainable societies.

The Rio definition leaves room for interpretation on when and how to apply the precautionary principle. Recognizing the need for a clear and consistent approach, the European Commission adopted a Communication in the year 2000 (European Commission 2000), which provides guidance on the use of the precautionary principle by

- Outlining the Commission's approach to using the precautionary principle
- Establishing guidelines for applying the precautionary principle

- Building a common understanding of how to assess, appraise, manage, and communicate risks that science is not yet able to evaluate fully
- Avoiding unwarranted recourse to the precautionary principle, which in certain cases could serve as a justification for disguised protectionism

WHO is also addressing decision making in environmental and health matters under scientific uncertainty and complexity. As part of the Health Impact Assessment Methods and Strategies Programme, the role and relevance of the precautionary principle in protecting human health are to be further clarified (WHO 2009).

The Fourth Ministerial Conference on Environment and Health (WHO Regional Office for Europe 2004), recognizing the Rio Declaration and the European Commission's Communication, reaffirmed the importance of the precautionary principle as a risk management tool. In the Conference Declaration, it is recommended that the precautionary principle should be applied "where the possibility of serious or irreversible damage to health or the environment has been identified and where scientific evaluation, based on available data, proves inconclusive for assessing the existence of risk ..." (WHO Regional Office for Europe 2004). The document *The Precautionary Principle: Protecting Public Health, the Environment and the Future of Our Children* was prepared as a background document (WHO 2004).

6.2.2.6.2 Methods for Predicting Impacts

All predictions in an EIA are based on conceptual models of the environmental systems. Several approaches and instruments can be used for predicting impacts. Each covers the range of impacts only partially and should therefore be used in conjunction with others:

- *Field and laboratory experimental methods*: This might include tests to predict impacts of a certain agent or activity on an indicator (e.g., a sensitive species), such as testing of desalination effluents in whole effluent toxicity tests using sensitive endemic species.
- *Physical or image models*: This involves the design and construction of small-scale models to study effects with a high degree of certainty in miniature.
- *Analog models*: Predictions are based on analogies (i.e., by comparing the potential impacts of the proposed desalination project with those of a similar, existing project).
- *Mathematical models*: Models based on cause–effect relationships are used, which vary in complexity from simple input–output relationships to highly sophisticated dynamic models with a wide range of interrelations, variable parameters, and coefficient constants that have to be identified and determined.
- *Mass balance models*: These models are based on the difference in the sum of the inputs compared with the sums of the outputs.
- *Matrices*: For predicting the impacts of a project, a two-dimensional matrix is often used, which cross-references the project activities on one axis with the environmental, socioeconomic, and human health setting in the project site on the other axis.

6.2.2.6.3 Criteria for Evaluating the Significance of Impacts

General criteria can be used to assess the significance of environmental and socio-economic impacts of a desalination project. These criteria are not mutually exclusive but are very much interrelated. The following general criteria should be taken into account when examining potentially significant adverse effects:

- Nature of impacts (direct/indirect, positive/negative, cumulative, transboundary)
- Time span (short/medium/long term, permanent/temporary, often/seldom)
- Extent (geographical area, size of affected population/habitat/species)
- Magnitude (low/severe, reversible/irreversible)
- Probability (high/medium/low probability, certain)
- Possibility to mitigate, avoid, or offset significant adverse impacts

6.2.2.7 Step 7: Mitigation of Negative Effects

The consideration of major alternatives, such as alternative location, technology, etc., should start early in the planning of a new project (cf. Step 2: Scoping), as the disposition to make major modifications is typically still high at this time. As project planning consolidates, major alternatives will be seriously considered only if the EIA has revealed significant impacts that cannot be mitigated otherwise. The investigation of impact mitigation measures should thus be understood as a process, which starts with the consideration of major alternatives in early project planning and continues after potential impacts have been analyzed (Step 6). At this stage, specific recommendations need to be elaborated that mitigate the predicted effects of the project.

The step of impact mitigation should identify the most feasible and cost-effective measures to avoid, minimize, or remedy significant negative impacts to levels acceptable to the regulatory agencies and the affected community. The definition of "acceptable" will vary according to different national, regional, or local standards, which depend on a society's or community's social, ideological, and cultural values, on economic potentials and on politics. For impacts that cannot be mitigated by technically and economically feasible methods, compensation methods should be identified. These may include monetary compensation or remediation activities.

The elements of mitigation (UNEP 2002) are organized into a hierarchy of actions:

- *Prevention*: Avoid impacts by preventive measures, consider feasible alternatives, and identify the best practicable environmental option.
- *Minimization*: Identify customized measures to minimize each of the main impacts predicted and ensure that they are appropriate, environmentally sound, and cost-effective.
- *Remediation*: Remedy or compensate for adverse residual impacts that are unavoidable and cannot be reduced further, as a last resort.

Mitigation can include *structural measures* (e.g., design or location changes, technical modifications, waste treatment) and *nonstructural measures* (e.g., economic incentives, policy instruments, provision of community services, capacity building). Remediation and compensation may involve *rehabilitation* of the affected site (e.g., habitat enhancement, restocking of fish), *restoration* of the affected site to its previous state after project demolition, and *replacement* of resource values at another location.

6.2.2.8 Step 8: Summary and Conclusions

In this step, the main findings and recommendations of steps 5–7 are summarized. The focus should be on the key information that is needed for drawing conclusions from the EIA results.

An overview of the main impacts (possibly in the form of a table) should be provided for this purpose, distinguishing between significant impacts that can be prevented or minimized and those that cannot. Direct and indirect impacts, positive and negative impacts, as well as potential cumulative effects should be considered.

Mitigation or alternative options should be offered for significant impacts where possible. In essence, the original project proposal should be systematically compared with alternative project configurations in terms of adverse and beneficial impacts and effectiveness of mitigation measures. As far as possible, trade-offs and uncertainties should be mentioned.

Finally, the "best practicable environmental option" should be identified, which is the preferred project configuration under environmental, social, cultural, and public health criteria. It should be ensured that this option is both economically and technologically feasible. The decision should be transparent and backed by arguments.

6.2.2.9 Step 9: Establishment of an Environmental Management Plan

An environmental management plan should be elaborated to ensure the ongoing assessment and review of the effects of the proposed desalination project during construction, commissioning, operation, maintenance, and decommissioning. It thus builds continuity into the EIA process and helps to minimize environmental impacts and optimize environmental benefits at each stage of project development. Attention should be given to involving the public in EIA implementation activities—for example, by establishing stakeholder monitoring committees.

The key objectives of EIA implementation and follow-up are to (UNEP 2002)

- Identify the actual environmental, socioeconomic, and public health impacts of the project and check if the observed impacts are within the levels predicted in the EIA.
- Determine that mitigation measures or other conditions attached to project approval (e.g., by legislation) are properly implemented and work effectively.
- Adapt the measures and conditions attached to project approval in the light of new information or take action to manage unanticipated impacts, if necessary.
- Ensure that the expected benefits of the project are being achieved and maximized.
- Gain information for improving similar projects and EIA practice in the future.

To achieve these objectives, the management plan should specify any arrangements for planned monitoring, surveillance, and/or auditing activities, including methodologies, schedules, protocols for impact management in the event of unforeseen events, etc. The main components and tools of EIA implementation and follow-up (UNEP 2002) as part of a management plan include

- *Monitoring activities*: Measure the environmental changes that can be attributed to project construction and operation, check the effectiveness of mitigation measures, and ensure that applicable regulatory standards and requirements are being met—for example, for waste discharges and pollutant emissions.
- *Surveillance activities*: Oversee adherence to and implementation of the terms and conditions of project approval.
- *Auditing activities*: Evaluate the implementation of terms and conditions, the accuracy of EIA predictions, the effectiveness of mitigation measures, and the compliance with regulatory requirements and standards.

6.2.2.10 Step 10: Review of the EIA and Decision-Making Process

The purpose of review is to verify the completeness and quality of the information gathered in an EIA. This final step will ensure that the information provided in the report complies with the ToR as defined during scoping (Step 2) and is sufficient for decision-making purposes. Review is a formal step in the EIA process and serves as a final check of the EIA report, which will then be submitted for project approval.

The review may be undertaken by the responsible authority itself, another governmental institution, or an independent body. Participation of collaborating and advisory agencies in the review process is strongly recommended, as is the involvement of the public and major stakeholders in public hearings about the outcomes of the EIA. The review should follow a systematic approach. This will entail an evaluation and validation of the EIA methodology and procedure and a check for consistency, plausibility, and completeness of the identified impacts, proposed alternatives, and suggested mitigation measures. The review process (UNEP 2002) can be based on explicit guidelines and criteria for review or may draw on general objectives, such as the following questions:

- Does the report address the ToR?
- Is the requested information provided for each major component of the EIA report?
- Is the information correct and technically sound?
- Have the views and concerns of affected and interested parties been considered?
- Is the statement of the key findings complete and satisfactory (e.g., for significant impacts, proposed mitigation measures)?
- Is the information clearly presented and understandable?
- Is the information sufficient for the purpose of decision making and condition setting?

The response to the last question is the most significant aspect for review and will largely determine whether or not the EIA can be submitted to the competent authority as it is or with minor revisions for decision making.

The competent authority will form its own judgment on the proposed project based on the EIA report, the analysis of stakeholder interests, statements from collaborating agencies, etc., and decide on approval or rejection of the proposed project. The competent authority will typically impose conditions if the project is approved, such as mitigation measures, limits for emissions, or environmental standards that must be observed.

6.3 SUMMARY AND RECOMMENDATIONS FOR EIA

An EIA is a procedure that identifies, evaluates, and develops means of mitigating potential impacts of proposed activities on the environment. Its main objective is to promote environmentally sound and sustainable development through the identification of appropriate mitigation measures and alternatives. Based on the EIA results, a decision must be reached that balances the positive and negative effects of a project in terms of their societal and environmental costs and benefits.

A formal EIA is often required for major development projects, such as a power plant or motorway, and may also be requested by authorities for projects in the water sector, such as for a dam and reservoir, water pipeline, wastewater treatment plant, or desalination facility. An EIA can also be adopted for plans, policies, or programs (e.g., water resources or coastal zone management plans) in the form of a strategic environmental assessment. The present chapter is primarily considered as guidance for impact assessment on a project level. Depending on the proposed project, it is incumbent upon the national authorities to define the scope and requirements for each EIA.

In general, an EIA should predict the impacts related directly or indirectly to the implementation of a desalination project. Where appropriate, this should comprise implications including ecosystem, socioeconomic, cultural, and public health effects, as well as cumulative and transboundary implications. The EIA should identify the positive effects and offer measures for mitigation of negative impacts.

The impact mitigation step should identify the most feasible and cost-effective measures to avoid, minimize, or remedy significant negative impacts to levels acceptable to the regulatory agencies and the affected community. The definition of "acceptable" will vary according to different national, regional, or local standards, which depend on a society's or a community's social, ideological, and cultural values, on economic potentials and on politics.

An EIA for a desalination project should address the following three areas of impact:

1. *Abiotic and biotic environment*: Abiotic factors include characteristic landscape and natural scenery, as well as soils and sediments, air, and water quality. The biotic environment encompasses the terrestrial and marine biological resources, including flora and fauna, in particular sensitive species, that inhabit the area impacted by the proposed project.
2. *Socioeconomic and cultural environment*: Socioeconomic and cultural considerations include the project's effects on the day-to-day lives of the

individuals and the community, the project's impact on the management of natural resources, and the project's impact on local and regional development. Gender- and age-specific effects and variations among the potentially affected population, such as social or ethnic affiliations, should be considered where appropriate.

3. *Public health*: Public health addresses the quality-of-life (well-being), improvement in community health, and potential risks associated directly or indirectly with the desalination project.

Key aspects that should be investigated in EIAs for desalination projects are

- Impacts related to the siting of the desalination facility and supporting infrastructure.
- Impacts related to the intake of source water, in particular, impacts due to the construction of intake structures and the entrainment and impingement of organisms.
- Impacts related to the discharge of concentrates, cooling water, and other waste streams resulting from the process, including an investigation of the hydrological mixing process in the discharge area; an investigation of impacts related to salinity, temperature, pH, or dissolved oxygen levels deviating from ambient conditions in the mixing zone; and an investigation of environmental fate and potential toxic effects of chemical residuals and by-products from the process, particularly of pretreatment and cleaning chemicals.

The EIA process proposed for desalination projects involves 10 basic steps:

1. Decide, on the basis of a screening process, whether or not an EIA is required.
2. Conduct scoping to determine the contents and extent of the EIA.
3. Identify policy and administrative aspects relevant to the project and the EIA.
4. Describe the technical design and process of the proposed desalination project.
5. Describe and assess the environmental baseline of the project site.
6. Describe and evaluate the potential impacts of the project on the environment.
7. Identify approaches for mitigation of negative impacts.
8. Provide a summary of the major findings, and develop conclusions.
9. Establish a program to monitor impacts during construction and operation.
10. Review the EIA process for decision-making purposes.

The following aspects should be considered when carrying out an EIA:

- *Precautionary principle*: Gaps of knowledge and uncertainties should be clearly identified and the precautionary principle, as defined by the UN, applied, as appropriate, in the EIA.
- *Public involvement*: Public involvement is an integral part of the planning and decision-making process and implementation of desalination projects for community water supply.

Additional recommendations that go beyond the scope of individual EIAs are the following:

- *Implementation of management plans*: To manage increasing desalination activity on a regional or national scale and to make it compatible with other human activities and nature conservation in the coastal zone, it is recommended that plans be elaborated that go beyond the scope of individual projects, in particular water resources management and integrated coastal zone management plans. Water resources management plans should cover a suite of supply and demand options, including water conservation programs and education, the use of water-saving devices, and water recycling for agricultural, industrial, and environmental applications.
- *Future research*: The present knowledge of the environmental, socioeconomic, cultural, and human health implications of desalination activity is limited. It is recommended that further research be initiated, monitoring of existing facilities be conducted, case studies be initiated, and results from monitoring, case studies, and EIAs be made available to a wider public to improve the understanding of the actual impacts of desalination activity on humans and the environment. To facilitate this process, the establishment of a clearinghouse for collecting and distributing the information should be considered.

REFERENCES

Cotruvo, J. 2009. Personal communication.

EU. 2001a. *Guidance on EIA. Screening*. Luxembourg: European Commission, Office for Official Publications of the European Communities. http://ec.europa.eu/environment/eia/eia-guidelines/g-screening-full-text.pdf.

EU. 2001b. *Guidance on EIA. Scoping*. Luxembourg: European Commission, Office for Official Publications of the European Communities. http://ec.europa.eu/environment/eia/eia-guidelines/g-scoping-full-text.pdf.

European Commission. 1997. *EIA directive 85/337/EEC on the assessment of the effects of certain public and private projects on the environment, amended by Directive 97/11/EC*. http://ec.europa.eu/environment/eia/eia-legalcontext.htm.

European Commission. 1998. *A guide to gender impact assessment*. Luxembourg: European Commission, Directorate-General for Employment, Industrial Relations and Social Affairs.

European Commission. 2000. *Communication from the commission on the precautionary principle*. COM (2000)1. http://ec.europa.eu/comm/environment/docum/20001_en.htm.

European Commission. 2001. *SEA directive 2001/42/EC on the assessment of the effects of certain plans and programmes on the environment*. http://ec.europa.eu/environment/eia/sea-legalcontext.htm.

Hassan, A.A., M.H. Birley, E. Giroult et al. 2005. *Environmental health impact assessment of development projects: A practical guide for the WHO Eastern Mediterranean Region*. Cairo: World Health Organization, Regional Office for the Eastern Mediterranean, Regional Centre for Environmental Health Activities. http://www.who.int/water_sanitation_health/resources/emroehiabook/en/index.html.

ICEW. 1992. *The Dublin statement on water and sustainable development.* Adopted January 31, 1992, in Dublin, Ireland. International Conference on Water and the Environment, Dublin 1992, organized by the United Nations World Meteorological Organization. http://www.un-documents.net/h2o-dub.htm.

UNECE. 1991. *Convention on environmental impact assessment in a transboundary context (Espoo, 1991).* United Nations Economic Commission for Europe. http://www.unece.org/env/eia/.

UNEP. 1992. *Rio declaration on environment and development.* Nairobi: United Nations Environment Programme. http://www.unep.org/Documents.Multilingual/Default.asp?documentID=78&articleID=1163.

UNEP. 2002. *Environmental impact assessment training resource manual.* 2nd ed. *Vols. 1 and 2.* Geneva: United Nations Environment Programme [print version available at Earthprint (http://www.earthprint.com); electronic version available at UNEP (http://www.unep.ch/etu/publications)].

UNEP. 2008. *Desalination resource and guidance manual for environmental impact assessments.* Manama: United Nations Environment Programme, Regional Office for West Asia; Cairo: World Health Organization, Regional Office for the Eastern Mediterranean. http://www.unep.org.bh/Publications/Type7.asp.

United Nations. 1997. *Report of the economic and social council for 1997.* A/52/3.18. September 1997. http://www.un.org/documents/ga/docs/52/plenary/a52-3.htm.

WHO. 1948. Preamble to the Constitution of the World Health Organization as adopted by the International Health Conference, New York, June 19–22, 1946, and entered into force on April 7, 1948. Geneva: World Health Organization.

WHO. 2004. *The precautionary principle: Protecting public health, the environment and the future of our children.* Copenhagen: World Health Organization, Regional Office for Europe. http://www.euro.who.int/document/e83079.pdf.

WHO. 2009. *Health impact assessment.* Geneva: World Health Organization. http://www.who.int/hia/en/.

WHO Regional Office for Europe. 2004. *Fourth ministerial conference on environment and health, Budapest, Hungary,* June 23–25, 2004. Declaration (EUR/04/5046267/6). http://www.euro.who.int/document/e83335.pdf.

World Bank. 1991. *Environmental assessment sourcebook. Vol. I (Policies, procedures, and cross-sectoral issues), Vol. II (Sectoral guidelines), Vol. III (Guidelines for environmental assessment of energy and industry projects)* [with updates]. Washington: World Bank. http://www-wds.worldbank.org/.

Index

D

E

N

O

P

Q

R

S